*EXPLORING QUANTUM
PHYSICS THROUGH
HANDS-ON PROJECTS*

EXPLORING QUANTUM PHYSICS THROUGH HANDS-ON PROJECTS

DAVID PRUTCHI AND SHANNI R. PRUTCHI

A JOHN WILEY & SONS, INC., PUBLICATION

Published by John Wiley & Sons, Inc., Hoboken, New Jersey
Published simultaneously in Canada

For general information on our other products and services or for technical support, please contact our Customer Care Department within the United States at (800) 762-2974, outside the United States at (317) 572-3993 or fax (317) 572-4002.

Wiley also publishes its books in variety of electronic formats. Some content that appears in print may not be available in electronic format. For more information about Wiley products, visit our web site at www.wiley.com.

Library of Congress Cataloging-in-Publication Data:

Prutchi, David.
Exploring Quantum Physics Through Hands-On Projects/David Prutchi and Shanni R. Prutchi.
  p. cm.
  Includes index.
  ISBN 978-1-118-14066-6 (pbk.)
1. Quantum theory—Popular works. 2. Quantum theory—Experiments. 3. Science projects.
I. Prutchi, Shanni R. II. Title.
  QC174.12.P785 2011
  535'.15—dc23   2011030360

10 9 8 7 6 5 4 3 2

In memory of Zeide Simon.
Dedicated to Saba Shlomo, Savta Ruthi, Babbe Rosmari,
Dorith, Hannah, and Abigail.

CONTENTS

INTRODUCTION

Tell me and I forget. Teach me and I remember. Involve me and I learn.
—Benjamin Franklin

Physics developed steadily after the introduction of Isaac Newton's ideas in the 1600s and had made great progress by the nineteenth century. People really felt the impact of this knowledge when the Industrial Revolution was made possible by the application of everything that scientists had learned about mechanical forces, gravity, electricity, magnetism, heat, light, and sound.

By the late nineteenth century, scientists felt that all of this understanding of physics formed a framework that could describe the world very deeply and thoroughly. Still, there were some nagging inconsistencies between theoretical calculations and experimental data, which were acknowledged by Lord Kelvin (who formulated the First and Second Laws of Thermodynamics) in his 1900 lecture titled "Nineteenth-Century Clouds over the Dynamical Theory of Heat and Light." The two "dark clouds" to which he was alluding were the unsatisfactory explanations that the physics of the time could give for the constancy of the speed of light, as well as for the glow produced by a hot body.

What is now known as "modern physics" was born from the two major physical theories that were developed during the twentieth century to resolve these two "dark clouds": for the former, the Theory of Relativity; for the latter, quantum mechanics.

The shift caused by modern physics was dramatic, because new concepts showed aspects of reality that are completely different from our day-to-day observations. The Theory of Relativity describes how events look very different for two observers who are moving relative to one another. It makes really bizarre predictions about the way clocks tick relative to one another when they move, how objects are measured to be shortened in the direction that they are moving with respect to the observer, and how energy and mass are equivalent (described by the famous equation $E = mc^2$).

Quantum mechanics makes statements that are even weirder than those of relativity. In the odd world of quantum physics—at least as explained through the extreme thought experiments of its original founders—objects can be in two places at the same time, cats can be both simultaneously dead and alive, and everything that looks continuous to us is really pixelated into tiny, discrete chunks.

Exploring Quantum Physics Through Hands-On Projects. By David Prutchi and Shanni R. Prutchi
© 2012 John Wiley & Sons, Inc. Published 2012 by John Wiley & Sons, Inc.

In spite of its strangeness, a working knowledge of relativity requires no more than basic algebra and geometry. After all, relativity is just a more fundamental, up-to-date, and accurate version of the classical physics founded by Newton. On the other hand, quantum mechanics is much more difficult to understand. Richard Feynman, a physicist who won the Nobel Prize for his work on quantum physics, didn't believe that anyone really understands what the theory tells us. In his words: "I think I can safely say that nobody understands Quantum Mechanics."

Nevertheless, quantum mechanics works so well that it has enabled the development of lasers, transistors, chips, and displays used in the electronic gadgets that are so important to our modern lives. Because of this, quantum physics ought to be an important part of everyone's education. However, the math is so complex, and our most intuitive notions about reality are so shockingly wrong, that understanding of this subject has remained largely confined to a select group of physicists and engineers. Unfortunately, this has also led to many popularizations that grossly misguide readers and give them completely false notions about the concepts and implications of quantum mechanics.

In this book we will try a different approach. The idea is to build an intuitive understanding of the principles behind quantum mechanics through hands-on construction and replication of the original experiments that led to our current view of the quantum world. We have developed the experimental setups in such a way that they can be constructed easily and at low cost. In addition, we have worked and re-worked the math, so that it is accessible to anyone with knowledge of high school algebra, basic trigonometry, and, if possible, a little bit of calculus. We want to point out that in spite of the many simplifications that we make, we strive to present a conceptually correct view of quantum physics to those who are not conversant in its highly specialized jargon and formalism.

Our approach comes from the belief that there is a huge difference between knowing about something and actually understanding it. We believe that if one is to understand anything about quantum mechanics, one must first develop a "gut feeling" about the quantum world, to get past the mystical veil that so tightly wraps its inner workings. We are hands-on tinkerers, so we follow Benjamin Franklin's approach toward education: "Tell me and I forget. Teach me and I remember. Involve me and I learn."

Our hope is that the do-it-yourself approach will demystify quantum physics, and help you navigate away from sensationalistic, speculative, or outright false accounts of this incredibly beautiful field.

Quantum effects tend to vanish as the size of an object increases. For this reason, quantum experiments are practical only for very small objects, such as photons (bits of light), electrons, protons, and other subatomic particles.* In this book, we will show you how to adapt commonly available electronic components, hardware store supplies, and other relatively low-cost items that can be purchased online to reproduce some of the most ground-breaking experiments ever done in physics!

---

*The difficulty of studying quantum effects on even slightly larger objects, such as atoms or molecules, is so great that only a few labs around the world are equipped to handle this difficult task.

Throughout the book, we will slowly build up a quantum picture of the world. The first chapter will have you become familiar with the way nineteenth century physicists understood light. Despite the advent of quantum mechanics, nineteenth century optics is still used today to make camera lenses, glasses, telescopes, and microscopes. However, the misbehavior of classical optics under very specific circumstances was one of the two clouds looming on Lord Kelvin's horizon. Understanding the classical view of light is key to appreciating why quantum mechanics would stir such a revolution in physics. In the first chapter, we will perform the experiments that seemed to confirm the correctness of the classical understanding of light's nature, but we'll also look at some of the problems raised by these same experiments.

In the second chapter, we will replicate the experiments that produced the data that could not be reconciled with the theoretical explanations of classical physics. We will study the sweeping explanations proposed by Max Planck and Albert Einstein to resolve this issue.

The third chapter will get us into atomic physics and radioactivity. We will build equipment to perform the experiments that gave us our current view of atoms and that brought chemistry into the modern era.

Next, in chapter 4, we will look at *quantization*—the core principle behind quantum mechanics—and at some of the ways in which it successfully tackles one of Lord Kelvin's "dark clouds."

In the experiments of chapter 5, we will take advantage of technology that was unavailable to the pioneers of quantum mechanics to show that both light and material objects can behave with the characteristics of both waves and particles. This is where quantum physics starts to get really weird, allowing particles to behave as if they are in two places simultaneously!

Chapter 6 will introduce Heisenberg's Uncertainty Principle—the concept that we cannot measure the exact position and momentum of an object at the same time. We will see that this is not due to imprecise measurements. Technology is advanced enough to hypothetically yield correct measurements. Rather, the blurring of these magnitudes is a fundamental property of nature with truly mind-boggling implications about our view of reality.

In chapter 7, we will talk about Dr. Schrödinger and his famous pet. We won't be conducting any experiments with dead or alive (or zombie) cats, but we will build some hands-on demonstrations that show Schrödinger's legendary thought experiment in action.

Last, chapter 8 looks at demonstrations of the existence of *entanglement*. This quantum property is so uncanny that it caused Einstein to mock it by calling it "spooky action at a distance." Entanglement was proven only in the 1980s, but its deep implications are already causing radical changes in the way in which we view our world. We will also look at technologies that are being developed as a consequence of understanding the role of information in quantum mechanics, and will end the book by peeking at how entanglement is quickly making strides into areas that until recently were purely the domain of science fiction, enabling quantum teleportation, unbreakable cryptography, and quantum computing.

PROLOGUE

YOUR QUANTUM PHYSICS LAB

This book assumes that you have some experience in electronic prototype construction. The circuits actually work, and the schematics are completely readable. It will be easy for you to understand them if you know some circuit design. However, the tested, modular circuits, components, and software are easy to use to build practical instrumentation, even if you view them as "black boxes," and do not explore their theoretical basis.

Of course, there are some basic electronic instruments that you will need in order to build, test, and use the equipment that we describe in this book. At the very least, you should equip your lab with the following:

- Soldering pen, pen rest with wet sponge, solder wire, and solder wick. Preferably, you should use a 70-W soldering station in which you can adjust the temperature between 220–480°C.

- Assorted tools, including a sharp diagonal wire cutter, wire stripper, various screwdrivers, hobby knife, needle-nose pliers, etc.

- Two handheld, autoranging, $4\frac{1}{2}$-digit, digital multimeters (DMMs). We suggest that you purchase two identical units. Tektronix and Fluke multimeters are recommended, but any of their $4\frac{1}{2}$-digit look-alikes (sold for around $35) are okay.

- High-voltage probe (>40 kV) compatible with your multimeter.

- Dual-channel, 60-MHz oscilloscope with FFT module. We recommend a second-hand Tektronix TDS210 with FFT module. However, there are many look-alikes that will work equally well. As a lower-cost alternative, you can consider connecting your PC to a USB dual-channel, 60-MHz oscilloscope adapter.

- Adjustable, metered, dual-output, linear, bench power supply. You want a unit capable of delivering two outputs adjustable between 0 and at least 20 VDC at 2 A each. A used Tektronix PS280 would be ideal.

The following are really helpful, but not absolutely necessary:

- High-voltage DC power supply. Ideally, you want to acquire a well-regulated power supply with selectable polarity (that is, you can choose whether the output referenced to ground will be positive or negative) and a range of 0–2,000 V with at least 1-mA current capability. A used HP 6516A would be great.

- Function generator. Doesn't need to be fancy. Something that will produce sine, triangular, and square waves at frequencies of up to at least 1 MHz will do.

- 10-MHz pulse generator. Nothing excessive. We recommend something like a Global Specialties model 4010.
- Spare PC with at least 2-GHz Pentium 4 processor and 2-GB RAM. You should install student editions of Excel and MATLAB to log, plot, and analyze the experimental data that you will obtain.

Since the prime subject in quantum physics is the photon (a particle of light), you will need a few lasers. Fortunately, laser pointers have dropped dramatically in price, placing in our hands wonderful, well-behaved "photon sources" that just a few years ago would have been out of the reach of anyone but the most advanced labs in the world. We recommend that you equip your lab with the following lasers:

- IR diode laser, 980-nm wavelength, 30-mW power.
- Helium-neon (HeNe) laser, 632-nm wavelength (red), 5-mW power.
- Red laser pointer, 670-nm wavelength, 5-mW power.
- Green laser pointer, 532-nm wavelength, 5-mW power.
- Violet laser, 405-nm wavelength, 100-mW power.

Many of the setups require mechanical construction. We have tried to keep this to a minimum by using off-the-shelf parts, but you will need to do some drilling and shaping. Most of the enclosures are made of aluminum, so you should have a handheld drill with a full set of drill bits. You will also need a good hacksaw, tin snips, and a nibbler, as well as an assortment of tools to tap holes, bend metal rods, and assemble parts together.

In addition, you should start a well-organized "junk box" to keep your electronic and optical components. We use plastic stacking shoe boxes that are neatly labeled to keep our parts organized. Useful parts that you may want to collect from old equipment or surplus sales include:

- Capacitors, resistors, inductors, and other passive electronic components.
- Aluminum boxes, power supplies, power adapters, power cords, fuses, rocker switches, panel lights, panel meters, prototyping boards, and other parts to build and enclose instruments.
- High-voltage diodes and capacitors from microwave ovens, old TVs, and CRT monitors.
- Strong magnets from damaged hard drives.
- Old oscilloscope CRT screens, even if the CRTs are dead.
- High-speed op-amps, comparators, and other linear ICs.
- Light-emitting diodes (LEDs) of different colors.
- Laser LEDs and laser modules from CD, DVD, and Blu ray™ players/ recorders, and laser pointers.
- Lenses and high-quality mirrors, such as those found in old cameras, binoculars, and projectors.

- High-voltage flyback transformers from old color TVs and CRT monitors.
- Items containing small amounts of radioactive materials such as ionization-type smoke detectors, old-style lantern mantles, and old luminescent watch hands.
- Polarizers from camera lenses, sunglasses, and 3D movie glasses.

Lastly, a word about where to find bargain components and instruments—storefront surplus stores (at least those dealing with electronics and science) are a disappearing breed. Most surplus is traded today on eBay® and other Internet auction sites (e.g., www.ebid.net, www.LabX.com, etc.) The best finds on eBay usually come from estate sales, as well as from people who specialize in buying surplus lots from the government or hi-tech companies that are going out of business. Before bidding on anything, check on the reputation of the seller, and read the item description very thoroughly. From our experience, we can tell you that "unable to test" most often means "broken and not repairable." Especially when buying online, *caveat emptor!*

A few interesting brick-and-mortar surplus stores are still around.* They have a presence on the Internet, but you will find the most interesting pieces by rummaging through their shelves. If you travel around, try to take a detour and visit the following:

- Apex Electronics in the vicinity of Los Angeles, California: You will find racks and racks full of electronics equipment, as well as a yard full of junk (including rocket parts). Plan to spend a full day browsing or leave empty-handed and confused!
- The Black Hole of Los Alamos in Los Alamos, New Mexico: It is similar in character to Apex, but is heavily loaded with atomic research surplus from the Los Alamos National Lab.
- Fair Radio Sales in Lima, Ohio: You will find racks and pallets full of unclassified surplus equipment. It is quite far from the major cities in Ohio, but the drive is worthwhile, since it contains many wonderful one-of-a-kind items that don't make it into their catalog.
- Murphy's Surplus Warehouse near San Diego, California: You will find some of the best surplus military equipment to be found! Definitely a worthwhile place to visit.
- Skycraft Parts & Surplus, very close to Disney in Orlando, Florida: Very cool store. Worthwhile visiting, especially if you have had it with Mickey Mouse and his friends! It is full of otherwise unobtainable stuff, much of it surplus directly from NASA's Kennedy Space Center.
- Surplus Sales of Nebraska in Omaha, Nebraska: This is a place that we haven't visited, but we often purchase very unique equipment from them via their Web site. We can just imagine what unadvertised jewels their warehouse may contain!

---

*For readers in Europe, Army Radio Sales in London, United Kingdom, has an excellent selection of useful items. Unfortunately however, they are a mail-only business. You cannot visit their warehouse.

Whether you are a student, hobbyist, or practicing engineer, we hope that this book will help you find how easy it is to understand the principles of quantum physics by building and experimenting with sophisticated setups at a small fraction of the comparable commercial cost.

For additional information, updated software, and more information on the projects detailed in this book please visit our Web site at: www.prutchi.com.

IMPORTANT DISCLAIMER AND WARNINGS

LEGAL DISCLAIMER

The projects in this book are presented solely for educational purposes. The construction of any and all experimental systems must be supervised by an engineer, experienced and skilled with respect to such subject matter and materials, who will assume full responsibility for the safe use of such systems.

The authors and publisher do not make any representations as to the completeness or the accuracy of the information contained herein, and disclaim any liability for damages or injuries, whether caused by or arising from the lack of completeness, inaccuracies of the information, misinterpretations of the directions, misapplication of the circuits and information, or otherwise. The authors and publisher expressly disclaim any implied warranties of merchantability and of fitness of use for any particular purpose, even if a particular purpose is indicated in the book.

References to manufacturers' products made in this book do not constitute an endorsement of these products, but are included for the purpose of illustration and clarification. It is not the authors' intent to make any technical data presented in this book supersede information provided by individual manufacturers. In the same way, any citation of government and industry regulations and standards that may be included in this book are solely for the purpose of reference and should not be used as a basis for design or testing.

Since some of the equipment and circuitry described in this book may relate to or be covered by U.S. or other patents, the authors disclaim any liability for the infringement of such patents by the making, using, or selling of such equipment or circuitry, and suggest that anyone interested in such projects seek proper legal counsel.

Finally, the authors and publisher are not responsible to the reader or third parties for any claim of special or consequential damages, in accordance to the previous disclaimer.

SAFETY AND GENERAL PRECAUTIONS

Many of the projects presented in this book involve power supplies that pose severe electrical shock hazards. In addition, some of the projects involve sources of laser, microwave, or ionizing radiation that may present hazards to the user, especially to sensitive tissues, such as those of the eyes. It must be stressed to the builder the need to exercise safety precautions involving proper handling, building, and labeling of potentially dangerous equipment. The builder and users of equipment described in this book assume full responsibility for the safe use of such devices.

Warnings Related to the Use of High-Voltage Power Supplies

High-voltage power supplies present a serious risk of personal injury if not used in a safe manner or by unqualified personnel. Needless to say, you need to exercise utmost care when conducting an experiment that uses high voltage. However, you also need to be careful after turning off the power supply, because many power supplies and the devices to which they connect can store energy, so that even with the unit unplugged from the wall a lethal hazard can still exist.

Always remove metal objects such as rings, jewelry, and watches before working with high voltage. Keep one hand in a pocket or closed behind your back and we recommend you wear rubber-soled shoes. Always prove to yourself that there is no voltage present anywhere in the high-voltage equipment on which you are working. Never rely on just one switch to power down a high-voltage supply. Turn the power switch off and disconnect the cord from the wall outlet. Be sure no one will inadvertently reconnect the power while you are working on the device. When working with power supplies of several hundred volts or higher, be especially aware that fully discharged capacitors can "self-charge" through the phenomenon of dielectric absorption.

Lastly, never work with high voltage alone, and never leave a high-voltage experiment unattended!

Warnings Related to the Use of Lasers

Lasers produce a very intense beam of light. Even low-power laser pointers can cause permanent damage if pointed directly at the eye! The coherence and low divergence of laser light means it can be focused by the eye into an extremely small spot on the retina, resulting in localized burning and permanent damage in just an instant. Certain wavelengths of laser light can cause cataracts. Infrared and UV lasers are particularly dangerous, since the body's "blink reflex," which can protect an eye from excessively bright light, works only if the light is visible.

Before turning on a laser, always be sure that it is pointed away from yourself and others. Never look directly into a laser, and never direct a laser at another person. Follow the same rules for direct reflections of laser light from reflective surfaces.

Warnings Related to Exposure to Ultraviolet Light

Ultraviolet light can cause permanent eye damage. Do not look directly at UV radiation, even for brief periods. If it is necessary to view a UV source, do so through UV-filtered glasses or goggles to avoid damage to the eyes. Take appropriate precautions with pets and other living organisms that might suffer injury or damage due to UV exposure.

In addition, please note that light from violet LEDs and lasers may contain substantial amounts of UV light that is absorbed largely in the lens of the eye and may cause cataracts.

Warnings Related to Microwave Exposure

Although the microwaves generated by the Gunnplexers that we will describe in the book are weak, the output is sufficiently concentrated that it could cause eye damage at very close range. Never look into the open end of a Gunnplexer while it operates at a distance under 50 cm.

Warnings Related to the Use of Sources of Ionizing Radiation

The radioactive sources recommended for use in the experiments described in this book are professionally manufactured, sealed sources that are exempt from U.S. Nuclear Regulatory Commission and state licensing. They present no special storage or disposal requirements. The activities of these sources are sufficient to conduct nuclear science experiments using standard Geiger–Müller (GM) counters or scintillation detectors, yet low enough not to present any radiation hazard. Nevertheless, we recommend using lead shields when shipping or storing multiple gamma sources to reduce radiation levels.

Make sure that the radioactive source disks are never breached or damaged. The major hazard with a breached sealed source is that radioactive materials could enter the body by inhalation, skin absorption, or ingestion. Immediately place a damaged source inside a zippered plastic bag and dispose of it according to the manufacturer's instructions.

Everyday items such as smoke detectors, old lantern mantles, and watches with radium-painted dials are radioactive sources that must be treated with respect. Most of these items are no longer manufactured, as exposure to the radioactive material was a health threat to the employees who made them. These everyday items were not designed for instructional or experimental use, and may therefore be hazardous when used for purposes other than originally intended.

Another potential source of radiation in the experiments described in this book comes from the use of vacuum electron tubes powered at over 15,000 V. Detectable levels of X-rays may be produced, depending on the conditions in which these are operated. Caution must be exercised by the experimenter to ensure adequate safety.

Warnings Related to the Use of Vacuum Tubes

Vacuum tubes, especially large ones, present a safety hazard if the tube breaks. Flying glass and electrodes can travel great distances when a tube implodes. This is a particular danger when large tubes, such as CRTs, are used. Treat all glassware under vacuum with respect. Safety glasses should be worn at all times to protect your eyes from flying glass should the glassware break and implode. Before each use, check all vacuum glassware for scratches, cracks, chips, or other mechanical defects that could lead to failure.

Warnings Related to the Use of Rare-Earth Magnets

Some of the projects in this book involve the use of rare-earth magnets (e.g., neodymium and samarium-cobalt magnets), which produce very intense magnetic fields at close distances. The chief hazard with these magnets is that they are strong enough to cause injuries to body parts pinched between two magnets, or a magnet and a metal surface. In addition, magnets allowed to get too near each other can strike each other with enough force to chip and shatter the brittle material, and the flying chips can cause injuries.

Warnings Related to the Use of Chemicals

Some of the experiments described in this book involve the use of hazardous chemicals. Please read the original supplier's MSDS (material safety data sheet), and make sure that you understand how to properly handle each material regarding its toxicity, health effects, first aid, reactivity, storage, disposal, protective equipment, and spill-handling procedures.

Warnings Related to the Authors' Attempts at Humor

Throughout this book, the authors sometimes attempt to use humor to explain certain concepts in a light-hearted manner. However, the authors are not professional comedians, and thus cannot assure the desired effect. Be assured, however, that the authors do not intend to offend any dead, alive, zombie, or otherwise undecided cats.

ACKNOWLEDGMENTS

We are deeply indebted to George J. Telecki, publisher at Wiley-Interscience, for giving us the opportunity to publish our work. We also thank Wiley's editorial staff, and the copy editor Mary Safford Curioli, for their invaluable help in preparing this book.

We would also like to thank and acknowledge the anonymous reviewers who provided precious suggestions and insights that helped refine the scope of this book.

We are especially grateful to Elana Resnick for working with us to verify the experimental setups and for helping us write the Instructor's Guide for this book.

Most of all, we want to thank Dorith (Mom), Hannah, and Abigail for their patience and encouragement while we fought over writing style and grammar, made a mess in the garage while building equipment, spoke physics at the dinner table, and got everyone irritated and irradiated along the way.

ABOUT THE AUTHORS

David Prutchi received his Ph.D. in Engineering from Tel-Aviv University in 1994, and then conducted postdoctoral research at Washington University. His area of expertise is the development of active implantable medical devices, and he is currently the Vice President of Engineering at Impulse Dynamics. He is an adept do-it-yourselfer, dedicated to bringing cutting-edge experimental physics within grasp of fellow science buffs.

Dr. Prutchi has published over 30 papers and holds over 70 patents. He is the lead author of the book *Design and Development of Medical Electronic Instrumentation: A Practical Perspective of the Design, Construction and Test of Medical Devices*, which was published by John Wiley & Sons.

Shanni R. Prutchi is a high-school junior at Jack M. Barrack Hebrew Academy in Bryn-Mawr, Pennsylvania. As an avid science and engineering enthusiast, she conducts research with her father in the areas of radio-astronomy and quantum physics.

The authors' Web site is www.prutchi.com

CHAPTER 1

LIGHT AS A WAVE

Before we get into quantum physics, let's understand the classical view of light. As early as 100 C.E., Ptolemy—a Roman citizen of Egypt—studied the properties of light, including reflection, refraction, and color. His work is considered the foundation of the field of optics. Ptolemy was intrigued by the way that light bends as it passes from air into water. Just drop a pencil into a glass of water and see for yourself!

As shown in Figure 1a, the pencil half under the water looks bent: light from the submerged part of the stick changes direction as it reaches the surface, creating the illusion of the bent stick. This effect is known as *refraction*, and the angle at which the light bends depends on a property of a material known as its *refractive index*.

In the 1600s, Dutch mathematician Willebrord Snellius figured out that the degree of refraction depends on the ratio of the two materials' different refractive indices. Most materials have a refractive index greater than 1, which means that as light enters the material from air, the angle of the ray in the material will become closer to perpendicular to the surface than it was before it entered. This is known as Snell's Law, which states that the ratio of the sines of the angles of incidence and refraction (θ_1, θ_2) is equal to the inverse ratio of the indices of refraction (n_1, n_2):

$$\frac{\sin \theta_1}{\sin \theta_2} = \frac{n_2}{n_1}$$

Try it out yourself with a small laser pointer! As shown in Figure 1b, partially fill a small aquarium with water. Disperse some milk in the water to make it a bit cloudy, which will make the laser beam visible. Use smoke from a smoldering match or candle to make the laser beam visible in the air above.

Measure the angles between the rays and a line perpendicular to the water surface. The refraction coefficient for water is approximately $n_2 = 1.333$, and for air is more or less $n_2 = 1$. Do your measurements match Snell's Law?

NEWTON'S VIEW: LIGHT CONSISTS OF PARTICLES

In 1704, Sir Isaac Newton proposed that light consists of little particles of mass. In his view, this could explain reflection, because an elastic, frictionless ball bounces off a

Exploring Quantum Physics Through Hands-On Projects. By David Prutchi and Shanni R. Prutchi
© 2012 John Wiley & Sons, Inc. Published 2012 by John Wiley & Sons, Inc.

(a)

(b)

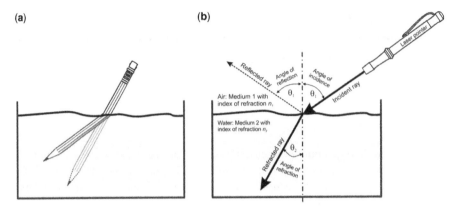

Figure 1   Refraction of light: **(a)** A pencil dipped in water appears distorted because refraction causes light to bend when it passes from one substance into another, in this case from air to water. **(b)** A laser pointer clearly demonstrates Snell's law of diffraction.

smooth surface just like light bounces off a mirror—that is, the angle of incidence equals the angle of reflection.

Remember that Newton was very interested in the way masses attract each other through the force of gravity. In his view, this force was responsible for refraction at the boundary between air and water. Newton imagined that matter is made up of particles of some kind, and that air would have a lower density of these particles than water. This is not far from what we know today—we would call Newton's particles "molecules" and "atoms." Newton then proposed that there would be an attractive force, similar to gravity, between the light particles and the matter particles.

Now, when a light particle travels within a medium, such as air or water, it is surrounded on all sides by the same number of matter particles. Newton explained that the attractive forces acting on a light particle would cancel each other out, allowing the light to travel in a straight line. However, near the air–water boundary, the light particle would feel more attracted by water than by air, given the water's higher density of "matter particles." Newton proposed that as the light particle moves into the water, it experiences an attractive force toward the water, which increases the light particle's velocity component in the direction of the water, but not in the direction parallel to the water.

This velocity increase in the direction perpendicular to the air–water boundary would deflect the light closer to perpendicular to the surface, which is exactly what is observed in experiments. Newton thus claimed that the velocity of light particles is different in different transparent materials, believing that light would travel faster in water than in air. (We now know this is not the case, but we'll get to that in a minute.)

Newton didn't equate gravity with the attractive force between matter particles and light particles. He needed this force to be equal for all light particles crossing the boundary between two materials to explain how a prism separates white light into the colors of the rainbow. Newton proposed that the mass of a light particle depended on its color. In his view, red light particles would be more massive than violet light particles. Because of their increased inertia, red light particles would thus be deflected less when crossing the boundary between materials.

Newton's greatness conferred credibility to his theory, but it was not the only one around. Dutch physicist Christiaan Huygens had proposed an earlier, competing theory: light consists of waves. This was supported by the observation that two intersecting beams of light did not bounce off each other as would be expected if they were composed of particles. However, Huygens could not explain color, and the wave versus particle debate for the nature of light raged until decisive experiments were carried out in the nineteenth century.

YOUNG'S INTERFERENCE OF LIGHT

Around 1801, Thomas Young discovered interference of light. This phenomenon is only possible with waves, providing conclusive evidence that light is a wave. In Young's experiments, light sent through two separate slits results in a pattern that is very similar to the one produced by the interference of water waves shown in Figure 2.

Let's spend some time experimenting with water waves before we go on to reproducing Young's experiments on the interference of light. Start by building a ripple tank, as shown in Figure 3, out of a glass baking pan (for example, a Pyrex® rectangular pan), some wood, two rubber bands, and a vibrating motor made for pagers and cellular phones.

The waves in the shallow layer of water are better observed by illuminating them from above to cast shadows through the glass bottom onto a white sheet of paper 50 cm below the tank. Use a spotlight, not a floodlight for illumination. Even better, use a strobe light (like the ones used by party DJs) to "freeze" the waves in place. Fill the pan with water to a depth of around 5 mm, and then fit pieces of metal sponge around the edges of the tank to reduce unwanted wave reflections from the pan's

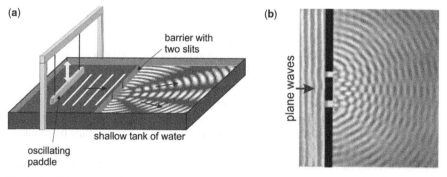

Figure 2   Water waves from two sources interfere with each other to form a characteristic pattern: **(a)** A ripple tank is a shallow glass tank of water used to demonstrate the basic properties of waves. In it, a shaking paddle produces waves that travel toward a barrier with two slits. **(b)** Plane waves strike two narrow gaps, each of which produces circular waves beyond the barrier, and the result is an interference pattern.

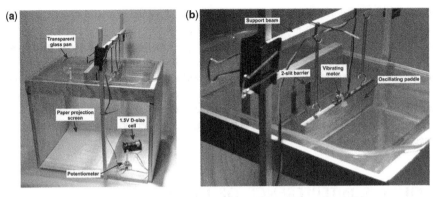

Figure 3   In our home-built ripple tank, a wooden stand supports a glass baking pan a distance away from a white sheet of paper. (a) With a light shining from above, ripples on a shallow layer of water in the pan are projected as shadows on the paper. (b) A small vibrating motor attached to a suspended beam just touching the water surface produces plane waves with which we can conduct experiments on wave reflection, refraction, and interference. For the sake of clarity, these pictures don't show the steel wool padding that we use to absorb reflections at the tank walls.

walls. Test the setup by dimming the room lights and lightly dipping a pencil into the water to create ripples.

To generate continuous plane waves, attach the vibrating motor to a wooden beam. Use rubber bands to suspend the beam from a support beam, and adjust the height of the vibrating beam so that it just touches the water surface. Power the motor from a 1.5-V D cell through a 100 Ω potentiometer (e.g., Clarostat 43C1-100).

Next, set up two straight barriers with a short one between them, along a line parallel to the vibrating beam. Make the gaps between barriers about 1-cm wide. Turn the potentiometer to generate straight waves with a wavelength of about 1 cm. Try different separations between the slits, and see if your data agree with the equation:

$$d = \frac{\lambda r}{s}$$

where d is the fringe separation (e.g., between the central fringe and the first fringe to its side), λ is wavelength, s is the distance between the slits, and r is the distance from the 2-slit barrier to the point where the fringes are observed.

Thomas Young did essentially the same thing using colored light instead of water waves. We will use inexpensive laser pointers and a simple double slit to replicate the experiment that Young performed to support the theory of the wave nature of light (Figure 4). Instead of making a double-slit slide,* we use one made by Industrial

---

*A double-slit slide can be made at home by coating a piece of clear glass with dark paint and then scoring the double slit with two narrowly-spaced razor blades. The best way to produce a quality slide is to apply two parallel strips of adhesive tape, leaving a $\frac{1}{2}$-in. band of glass uncovered. A large drop of paint applied toward one end of the bare strip is then spread with a razor blade along the strip to deposit a very smooth, constant-thickness layer of paint. Two brand-new razor blades should then be stacked, using a paper spacer between them. The two parallel blades should then be used with a brisk motion to score a pair of lines across the dry paint. The result should be two hairline transparent slits separated by an extremely thin line of paint.

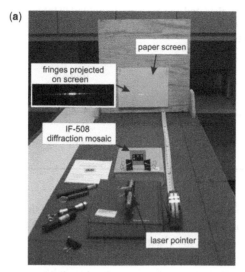

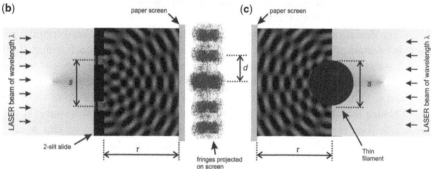

Figure 4   A modern version of Young's experiment to demonstrate the wave nature of light. (a) All that is needed is an inexpensive laser pointer and a slide with two slits. (b) The separation between fringes is related to the distance between the slits according to $d = \lambda r/s$. (c) A thin filament of thickness s produces the same interference pattern as a double slit of the same separation.

Fiber-optics. Their model IF-508 diffraction mosaic is a low-cost ($6) precision array of double slits and gratings for performing laser double- and multiple-slit diffraction experiments. The mosaic is mounted in a 35-mm slide holder and contains four double slits and three multiple-slit arrays on an opaque background with clear apertures. Double-slit separations range from 45 to 100 μm in width. The gratings are 25, 50, and 100 lines/mm.

Interestingly, Young found that the separation between fringes is related to the distance between the slits exactly through the same equation as the water analog:

$$d = \frac{\lambda r}{s}$$

TABLE 1   Calculated and Measured Double-Slit Interference Distance Between Center and First Bright Fringe for $s = 45$ μm at Different Wavelengths

Laser pointer	Wavelength λ [nm]	Measured fringe separation d [mm]	Calculated fringe separation d [mm]
Red	630	13.8	14
Green	532	11.8	11.8
Violet	405	8.6	9

where s is the distance between slits, λ is the wavelength of the light, d is the separation between fringes (the distance between central maximum and each of the first bright fringes to its side), and r is the distance from the slits to the screen.

For the double slit marked "25 × 25" in the IF-508 diffraction mosaic, $s = 45$ μm. Using red, green, and violet laser pointers with $r = 1$ m, we measured the fringe separations shown in Table 1.

The deviation between measured fringe separation and calculated fringe separation is because of our assessment of the location of the center of each fringe. Better accuracy can be obtained by repeated measurement and averaging. Try out this and other slit separations available in the IF-508 slide ($s = 5.8, 7.5$, and 10 μm) for yourself, and see how well the wave model accounts for the behavior of light.

Notice that $d = \lambda r/s$ is *not* dependent on the width of the slits, only on their separation. Interestingly, this same equation works when there are no slits at all. If one shines a laser pointer at a human hair in a dark room, the separation d between the interference fringes can be calculated by making s equal to the diameter of the hair.[1] Try it out! Shine a laser pointer at a hair and measure the distance between the interference fringes. Try to calculate the width of the hair—you should come up with a thickness of around 50 to 150 μm. Try to remember this, because the equivalence between a double-slit interference pattern and that obtained using a very thin filament will become very important in experiments that we will conduct later to expose quantum effects.

AUTOMATIC SCANNING OF INTERFERENCE PATTERNS

Accurately measuring interference patterns from projections on a screen is rather tedious. However, you can build a simple device that makes it possible to display interference patterns on an oscilloscope, making it easy to measure not only the distance between fringes, but also their amplitude.

As shown in Figure 5, the idea is to use a rotating mirror and a fast-light sensor to convert the interference pattern into an equivalent time-domain signal that can be displayed by a conventional oscilloscope. For the light sensor, we used a TAOS

TSL254R-LF light-to-voltage converter. This device is an inexpensive component that incorporates a light-sensitive diode and amplifier on a single chip. It is very easy to use. It requires a supply voltage in the range of 2.7 to 5.5 V (we use two 1.5-V AA batteries in series), and produces an output voltage that is directly proportional to the light intensity. We placed the light sensor behind a narrow slit built from two single-edge razor blades.

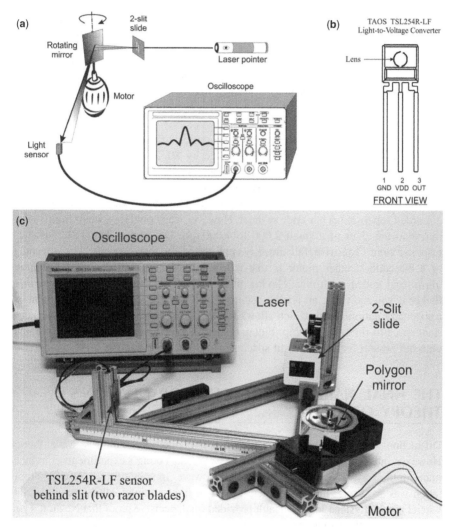

Figure 5   A simple scanner makes it easy to measure interference and diffraction patterns with an oscilloscope. (**a**) Simplified diagram of the basic concept. A small DC motor spins a mirror to scan the pattern onto a narrow-view light sensor, transforming the pattern's distribution along space into a signal that varies with time. An oscilloscope synchronized to the motor displays the pattern. (**b**) For the light sensor we used a TAOS TSL254R-LF light-to-voltage converter placed behind a narrow slit made from two razor blades. (**c**) We used a motor and polygon mirror from a broken bar-code scanner to build our setup.

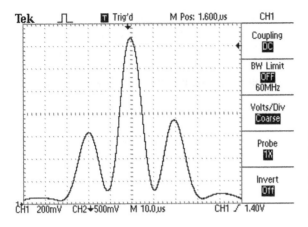

Figure 6   Interference pattern obtained with our scanner (Figure 5) for a double slit of $s = 10$-μm illuminated by a red (630-nm) laser pointer.

As shown in Figure 5c, we built the optical stand from 1-in. × 1-in. cross-section, T-slotted aluminum extrusions made by 80/20, Inc. These are meant for building office cubicles and machine frames, so they are widely available (e.g., from McMaster-Carr) and inexpensive. In spite of this, they are very rugged and sufficiently straight to perform optical experiments. Our motor and mirror came from a discarded supermarket bar-code scanner. However, you could rig a small front-surface mirror to the shaft of a small 2,000 to 4,000 rpm DC motor. The TSL254R-LF's response time (2 μs rise/fall time) is appropriate for these speeds. The advantage of a bar-code scanner motor is that it usually comes installed with a polygonal mirror and speed controller. Having more mirror surfaces per revolution reduces flicker if you are using an analog oscilloscope. The integrated controller maintains a constant rotation speed, which allows you to calibrate the system to produce a constant space-to-time relationship. Figure 6 shows a typical oscilloscope trace obtained with our system for a 10-μm slit spacing with a 630-nm red laser.

THE FINAL NAIL IN THE COFFIN FOR NEWTON'S THEORY OF LIGHT

Diffraction, reflection, and color are also explained by Young's wave theory. However, interference is the calling card of waves, so Young's experiments convinced many in the early 1800s that light is indeed a wave. In spite of this, Newton's reputation was so strong, that his particle model of light retained adherents until 1850, when French physicist Jean Foucault provided final, decisive proof that Newton's particle theory of light must be wrong. Remember that Newton's theory required the speed of light to be higher in water than in air? Well, Foucault experimentally showed the exact opposite. As shown in Figure 7, Foucault used a steam turbine to spin a mirror at the rate of 800 rps. He bounced a light beam off the rotating mirror; the beam was then reflected by a stationary mirror 9 m away. By the time the light returned to the rotating mirror, the mirror had rotated a little, causing the light to be deflected a certain amount away from the source.

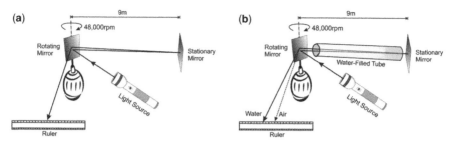

Figure 7　In 1850, Jean Foucault used this setup to measure the speed of light in (a) air and (b) water. He found that light travels more slowly in water than in air, contrary to the prediction of Newton's particle theory of light.

Foucault then placed a water-filled tube with transparent windows along the light path between the mirrors. If, as Newton affirmed, light travels faster in water than in air, the deflection angle would be smaller and the beam would arrive closer to the source.[†] Instead, Foucault found that introducing water in the optical path further delayed the beam, indicating that light travels more slowly in water than in air, contrary to the prediction of Newton's particle theory of light.

LIGHT AS AN ELECTROMAGNETIC WAVE

Later, in the 1860s, Scottish physicist James Clerk Maxwell identified light as an electromagnetic wave. Maxwell had derived a wave form of the electric and magnetic equations, revealing a wave-like nature of electric and magnetic fields that vary with time.

Maxwell figured out that an electric field that varies along space generates a magnetic field that varies in time and vice versa. For that reason, as an oscillating electric field generates an oscillating magnetic field, the magnetic field in turn generates an oscillating electric field, and so on. Together, these oscillating fields form the electromagnetic wave shown in Figure 8. The way in which an electromagnetic wave travels through space is described by its wavelength λ, while its oscillation in time is described by the wave's frequency. The frequency f and the wavelength are related through $c = \lambda f$, where c *is* the speed of light.

Because the speed of Maxwell's electromagnetic waves predicted by the wave equation coincided with the measured speed of light, Maxwell concluded that light itself must be an electromagnetic wave. This fact was later confirmed experimentally by Heinrich Hertz in 1887. Today, we use the electromagnetic spectrum at all wavelengths—from the enormously long waves that we use to transmit AC power, through the radio wavelengths that are the foundation of our wireless society, to the extremely short wavelengths of gamma radiation (Figure 9).

[†]A modern replication of Foucault's experiment is described in reference 2.

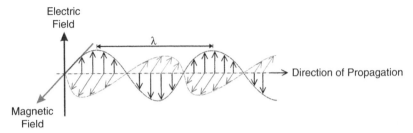

Figure 8   An oscillating electric field generates an oscillating magnetic field; the magnetic field in turn generates an oscillating electric field, and so on. Together these oscillating fields form an electromagnetic wave with wavelength λ that propagates at the speed of light c.

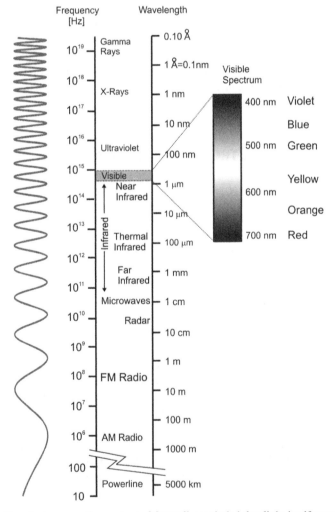

Figure 9   The electromagnetic spectrum. Maxwell concluded that light itself must be an electromagnetic wave. This fact was later confirmed experimentally by Hertz in 1887.

We understand that the only difference between visible light and the rest of the spectrum is that it is the range of electromagnetic waves to which our eyes are sensitive.

POLARIZATION

Polarization is an important characteristic of light that Maxwell's electromagnetic theory was finally able to explain. Notice in Figure 8 that the electric field is shown to oscillate in one plane, while the magnetic field oscillates on a perpendicular plane. The wave travels along the line formed by the intersection of those planes. The electromagnetic wave shown in this figure is said to be "vertically polarized," because the electric field oscillates vertically in the frame of reference we have chosen.

Light from most natural sources contains waves with electric fields oriented at random angles around its direction of travel. A wave of a specific polarization can be obtained from randomly polarized light by using a *polarizer*.

A polarizer can be made of an array of very fine wires arranged parallel to one another. The metal wires offer high conductivity for electric fields parallel to the wires, essentially "shortening them out" and producing heat. Because of the nonconducting spaces between the wires, no current can flow perpendicularly to them. As such, electric fields perpendicular to the wires can pass unimpeded. In other words, the wire grid, when placed in a randomly-polarized beam, drains the energy out of one component of the electric field and lets its perpendicular component pass with no attenuation at all. Thus, the light emerging from the polarizer has an electric field that vibrates in a direction perpendicular to the wires.

Although the wire-grid polarizer is easy to understand, it is useful only up to certain frequencies, because the wires have to be a fraction of the wavelength apart. This is difficult and expensive to do for short wavelengths, such as those of visible light. In 1938, E. H. Land invented the H-Polaroid sheet, which acts as a chemical version of the wire grid. Instead of long thin wires, it uses long thin polyvinyl alcohol molecules that contain many iodine atoms. These long, straight molecules are aligned almost perfectly parallel to one another. Because of the conductivity provided by the iodine atoms, the electric vibration component parallel to the molecules is absorbed. The component perpendicular to the molecules passes on through with little absorption.

As you will see throughout this book, understanding polarization is very important when experimenting with quantum physics, so we would like for you to gain an intuitive feel for this interesting property of waves.

OPTICS WITH 3-cm WAVELENGTH "LIGHT"

Let's start by experimenting with a polarizer that is actually made out of wires, such as the one shown in Figure 10. However, we'll need a source of electromagnetic waves with sufficiently large wavelength. Fortunately, it is easy to generate and detect

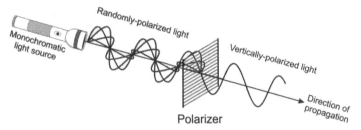

Figure 10   A parallel-wire polarizer absorbs electric field lines that are parallel to the wires. Only the perpendicular electrical field component of light is allowed to pass, producing light that is polarized perpendicularly to the direction of the wires.

microwaves with a wavelength of around 3 cm, making it possible to experiment with "optical" components scaled up to very convenient dimensions. Using a 3-cm microwave wavelength transforms the scale of the experiment. Measurements that would require specialized equipment at optical wavelengths to deal with submicrometer dimensions are easily accomplished with a simple ruler at 3-cm wavelengths.

As shown in Figure 11, a simple microwave transmitter can be built using a Gunnplexer,[3,4] which is a self-contained microwave module based on a specialized diode invented by John B. Gunn in the early 1960s. When a DC voltage is applied to the Gunn diode, current flows through it in bursts at regular intervals in the 10- to 100-GHz (10^{10} to 10^{11} Hz) range. These oscillations cause a wave to be radiated from the Gunnplexer's output slot.

You can find a Gunnplexer to use by taking apart a surplus microwave door opener or speed radar gun. The typical power output of Gunnplexers for these

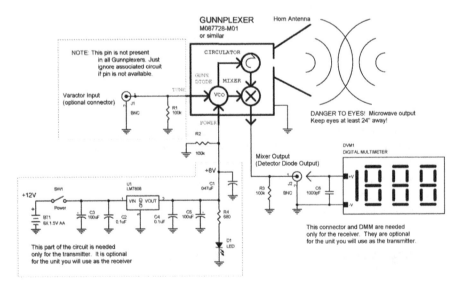

Figure 11   Schematic diagram for the Gunnplexer microwave transmitter/receiver. Two identical units can be built, but one simplified transmitter and one simplified receiver can also be used in these experiments.

applications is in the 5- to 10-mW range, and they commonly operate in either the so-called X-band (at 10.5 GHz) or K-band (24.15 GHz). For the receiver, you will need a second Gunnplexer built to operate in the same frequency range as your transmitter Gunnplexer, but this time you will use the microwave detector diode that is part of these modules.

As shown in Figure 12, we used surplus MO87728-M01 Gunnplexers, but almost any other model should work just as well. Aluminum die-cast boxes made by Bud Industries (model AN-1317) made nice enclosures for the transceivers. We bought the metallized-plastic horn antennas from Advanced Receiver Research.

A word of caution regarding the use of Gunnplexers: although the microwaves generated by Gunnplexers will not cook you, the output is sufficiently concentrated that it could cause eye damage at very close range. It is wise to never look at close range into the open end of a Gunnplexer while it operates.

Now to our experiments. Place the Gunnplexers about a meter apart and point the antennas at each other. Connect a digital voltmeter to the detector diode of your receiving Gunnplexer (the "mixer" output). Turn on the transmitter. The highest voltage across the mixer diode should appear when the Gunnplexers are oriented in the same plane. This is because Gunnplexers are polarized transmitters and receivers of microwaves. The electric field of the transmitted wave oscillates in the same orientation as the Gunn diode, and the detector is sensitive to fields in the same orientation

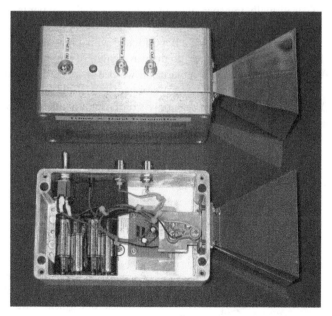

Figure 12   These are the X-band 10.5-GHz transmitter/receivers that we built from surplus Gunnplexer modules. Polarized microwaves with a wavelength of approximately 3 cm are launched from the horn antenna when the Gunn diode is powered. The "Mixer" diode in a second Gunnplexer is used to detect microwaves. It produces an output voltage proportional to the intensity of a properly polarized microwave signal.

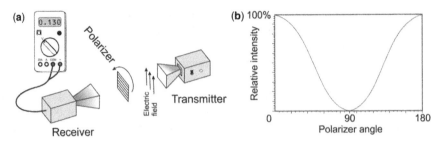

Figure 13 Experimenting with a polarizer. **(a)** Rotate a polarizer between two Gunnplexers that face each other. **(b)** Charting the voltmeter's measurements versus the polarizer angle should result in a graph similar to this one.

as the mixer diode. In our setup, we measure around 0.8 V output from the receiver when the horn antennas are placed right against each other. The signal drops down to 40 mV at a distance of 65 cm. You may note that the output voltage from the detector is negative with respect to ground. This is normal, and happens because of the way in which the mixer diode is internally connected within the Gunnplexer.

Next, you can build a polarizer by arranging copper wires in an array, just as in the idealized diagram of Figure 10. However, it is easier to use a circuit board made for prototyping—known as a stripboard—that already has conductors in an arrangement like the one we need. These boards are manufactured on an epoxy substrate that, fortunately for us, is virtually transparent to microwaves. Parallel copper tracks run along the board for hardwiring electronic components. These will act as the parallel wires for our polarizer. The whole board is usually perforated with a hole matrix, but the aperture of the holes is so small compared to the wavelength of the microwaves that they have no effect. For our microwave polarizers, we use stripboards manufactured by Vero. The specific one we use is the Veroboard™ 01-0021, which is sold for around $10 each by many electronics supply stores, but any other stripboard should work just as well.

Take a 10-cm × 12-cm piece of stripboard and place it between the transmitting and the receiving Gunnplexers, as shown in Figure 13a. Knowing that the electric field

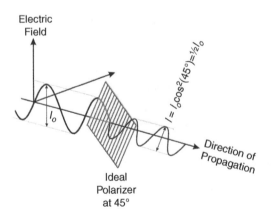

Figure 14 A polarizer actually shifts the polarization of an electromagnetic wave. The intensity of the exiting wave I is given by $I = I_o \cos^2 \theta_i$, where I_o is the initial intensity, and θ_i is the angle between the wave's initial polarization direction and the axis of the polarizer.

of the transmitted wave oscillates in the same orientation as the Gunn diode, can you predict how you should orient the polarizer to obtain the highest reading from the receiver? Using a protractor, graph the received intensity as you rotate the polarizer. You should end up with a graph that looks like the one in Figure 13b.

Next, take the polarizer out of the way and rotate the transmitter 90°. The signal at the receiver should drop close to zero. This makes sense, right? After all, the sensitive orientation of the receiver's detector diode is orthogonal to the electric field of the waves produced by the transmitter. Insert the polarizer at 0° and 90° referenced to the transmitter's polarization. The receiver still shows no signal. No surprise there . . . Now, rotate the polarizer to 45°. You should suddenly detect a signal. How can inserting the polarizer increase the signal level at the detector?

Look at Figure 14. Placing a polarizer at 45° introduces a component of the wave parallel to the receiver's sensitive axis, so that some of the transmitted signal is detected. An ideal polarizer in this case would allow half of the signal intensity to go through, but the signal exiting the polarizer would be rotated to 45°. At this new

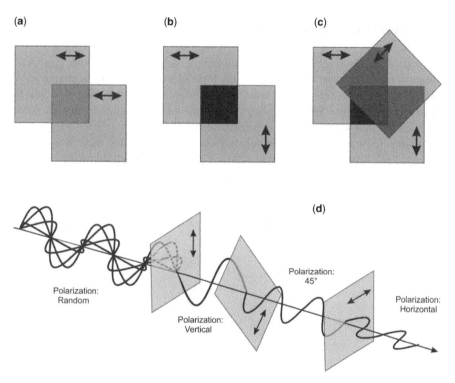

Figure 15   The intensity of light is cut by at least one-half when randomly polarized light is viewed through an ideal polarizer. (a) In practice, adding a second polarizer with the same orientation further attenuates the light, because real polarizers are not perfectly transparent. (b) Rotating one of the polarizers by 90° blocks virtually all the light from coming through. (c) However, inserting a 45° polarizer between the orthogonally oriented polarizers changes the intermediate polarization to 45°, allowing some light to emerge from the final polarizer (d).

polarization, the detector is able to pick up some signal, as you found out in the Figure 13 experiment.

The same exact effects can be studied using polarizing films and visible light. Although 3-cm microwaves are more than four orders of magnitude larger than visible light, they are both electromagnetic radiation and behave in the same way. Polarizing films are very inexpensive, so they allow easy and affordable experimentation with the important concept of polarization. Low-cost, linear-polarizing film is sold by educational and scientific supply companies such as Anchor Optics or Spectrum Scientifics.

Polarized film doesn't usually come labeled for its axis of polarization. However, the axis of polarization is easy to find by looking at sunlight reflecting from water. The glare is almost completely horizontally polarized (although this does depend on the height of the sun). The glare should be minimal when viewed by a vertically oriented polarizer.

Try out the experiment shown in Figure 15. Play around with the film and really try to build an instinctive understanding of polarization and polarizers.

REAL-WORLD BEHAVIORS

As you collect data, remember that Gunnplexers, stripboards, and other microwave components do not behave exactly as do their ideal, theoretical counterparts. For example, data from real Gunnplexers approximate, but do not exactly lie on, the ideal curve of Figure 13b. You should also have observed that stripboards are far from ideal polarizers, since they attenuate the microwave beam, even if they are perfectly oriented with respect to the polarization of the transmitter/receiver system. In addition, real-world polarizers also reflect a part of the beam, allowing "standing waves" to be formed at certain positions within the path between the Gunnplexers.

Conduct the experiment shown in Figure 16. Place the transmitter and receiver horns end-to-end, and then *slowly* move the units apart. Since the microwave horns are not perfect transmitters or receivers, they act as partial reflectors. As such, some of the

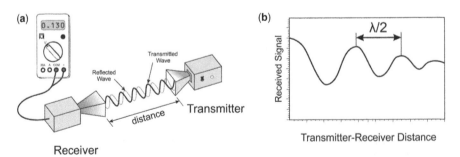

Figure 16   The microwave horns act as partial reflectors, so that the radiation from the transmitter reflects back and forth between the transmitter and the receiver **(a)**. **(b)** The transmitted and reflected waves will be in phase whenever the distance between the transmitter and the receiver is an integer multiple of $\lambda/2$, peaking the signal at the receiver.

radiation from the transmitter reflects back and forth between the transmitter and the reflector. However, if the distance between the transmitter and the receiver is an integer multiple of $\lambda/2$, then all the multiply-reflected waves entering the receiver horn will be in phase with the primary transmitted wave. Under this condition, the primary transmitted and reflected waves will be in phase, peaking the received signal.

Record the distance every time you find a maximum signal as you slide the units apart. How well are you able to calculate the signal's wavelength and frequency using this method? Compare it to the specified operating frequency of your Gunnplexers (X-band Gunnplexers commonly found in microwave door openers and radar speed guns operate at a frequency of 10.525 GHz, which has a wavelength of 2.85 cm.)

Play with your Gunnplexer setup so that you uncover its idiosyncrasies. For example, see how placing your hand close, but not in the way of the beam, affects your readings. What about sweeping the receiver away from the direct line between the antennas facing each other? Go ahead—try it out! You will find out that the microwave beam is not a pencil-tight beam like that of a laser pointer, but rather fits a rather wide Gaussian distribution.

Understanding the equipment is always an important step in the design of an experiment, since real-world components rarely behave in the same way as their theoretical counterparts. The various glitches and artifacts in data due to real-world behaviors in your equipment may completely obscure the effect that you are trying to measure. Even worse, sometimes these peculiarities trick you into believing that they are the very signal that you are trying to measure.

DOUBLE-SLIT INTERFERENCE WITH MICROWAVES

A double-slit experiment with 3-cm microwaves will give you the basic understanding of how double-slit experiments are conducted in sophisticated quantum research that we will discuss in chapters 7 & 8. Just as so many of the other specialized particle detectors, the Gunnplexer receiver does not produce an image, so the interference pattern needs to be measured one point in space at a time. Observing the double-slit interference with the Gunnplexers requires the receiver to be scanned along the plane where the interference pattern is formed.

You will need to build a *goniometer stand* to conduct this experiment. The stand has two guide rails joined by a pivoting joint, with a protractor to measure the angle between the arms. We built ours (Figure 17) from 1-in. × 1-in. cross-section, T-slotted aluminum extrusions made by 80/20, Inc. We joined two 60-cm-long extrusions using a digital protractor made by iGaging. These are commonly sold by woodworking tool suppliers to enable accurate setting of table-saw blade and fence angles. We built two Gunnplexer unit holders from 80/20, Inc. L-brackets and 4-hole plates. We tapped a $\frac{1}{4}$-in. hole in the back of the boxes we used to enclose the Gunnplexers, and used a knob screw to hold the units at any angle we require. A stick-on metric tape and two stick-on protractor dials complete our goniometer mount.

Make a slide holder out of sheet metal or a thin metal sheet that you can easily bend. Place a strip of adhesive-backed magnetic tape where the various slits and slides will be mounted. The plane for slide mounting should be centered on the

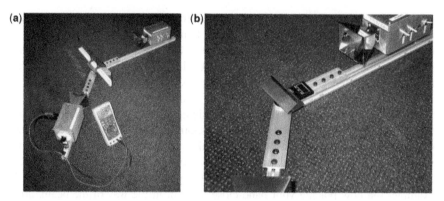

Figure 17   Our homemade goniometer stand is built from two pieces of 80/20, Inc. aluminum T-slot profile joined by a digital protractor. (a) Stick-on rulers and protractor dials make it easy to position the Gunnplexer units. (b) A folded sheet-metal holder with magnetic-tape backing supports slides right over the pivot point.

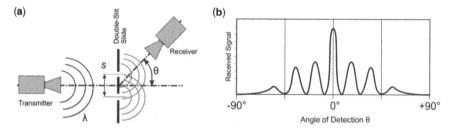

Figure 18   Many experiments to observe the quantum behavior of particles use a nonimaging detector mounted on a goniometer stand to measure the interference pattern. (a) Measuring the double-slit interference with microwaves is an excellent way to prepare for more advanced experiments. (b) With this geometry, peaks in the interference pattern are found at integer multiples of $s(\sin \theta) = \lambda$.

protractor's axis. Cut two double-slit slides out of sheet metal (or any other metal that sticks to a magnet). Make one spacer have a 6-cm-wide slit spacer, and the other a 9-cm-wide slit spacer. Make the slits 1.5-cm wide. Use a wooden dowel to keep the slotted slide straight.

Scan the receiver around the double-slit slide as shown in Figure 18a. You should find that the location of interference fringes as a function of the angle of detection is dependent only on the wavelength λ, and the center-to-center distance between the slits s. As the goniometer arms are moved, the received signal should peak at integer multiples of $s(\sin \theta) = \lambda$, as shown in Figure 18b.

THE DOPPLER EFFECT

Lastly, if you built at least one of the units to act as both a transmitter and receiver, you essentially have the heart of a police radar gun, and you may want to experiment with

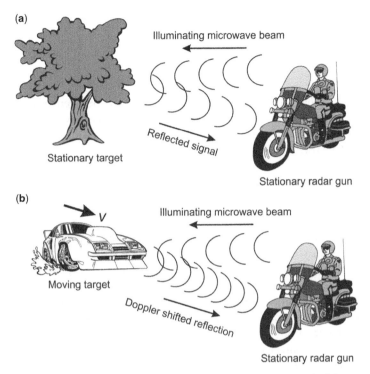

Figure 19   A Doppler radar illuminates its target with a microwave signal of frequency f_t. **(a)** A stationary object simply reflects the signal at the same frequency. However, a moving target shifts the reflected signal by $f_{\text{Doppler}} \approx 2V f_t/c$, where V is the relative velocity between the moving target and the radar gun, and c is the speed of light.

the interesting wave property of light called the Doppler effect. Connect the Gunnplexer's detector diode output (mixer output) to an audio amplifier, such as the microphone input of an old cassette tape recorder.[‡] It is a good idea to place a 1-μF nonpolar capacitor between the Gunnplexer's output and the audio amplifier's input to get rid of any DC currents. Turn on the transmitter and point it at a passing car. The whooshing sound that you will hear is the Doppler shift caused on the reflected microwaves by the car's movement.

Take a look at Figure 19. The signal reflected by a stationary object illuminated by microwaves from the policeman's stationary radar gun is of the same frequency as the source. No "beat" tone is generated when the transmitted and received frequencies are subtracted by the Gunnplexer's mixer diode. However, a moving object reflects a signal with frequency shifted in proportion to its speed relative to the stationary source. In this case, the mixer diode within the Gunnplexer reproduces the

[‡]A convenient, self-contained high-gain amplifier and speaker is sold by RadioShack for around $15 as their model number 277-1008.

Doppler shift, and a microcomputer inside the radar gun can calculate the target's speed using the approximation:

$$f_{\text{Doppler}} \approx 2V\frac{f_t}{c}$$

where f_{Doppler} is the Doppler shift, V is the relative velocity between the moving target and the radar gun, f_t is the frequency of the illuminating microwave beam, and c is the speed of light. When using a 10.5-GHz Gunnplexer, the Doppler shift for a car moving at 65 MPH ($29.0576 \frac{\text{m}}{\text{s}}$) will be:

$$f_{\text{Doppler}} = 2V\frac{f_t}{c} = 2 \times 29.0576 \left[\frac{\text{m}}{\text{s}}\right] \times \frac{10.5 \times 10^9 \, [\text{Hz}]}{299{,}792{,}458\left[\frac{\text{m}}{\text{s}}\right]} = 2035 \text{ Hz}$$

which is well within a human's hearing range. For each mile per hour of target speed, the Doppler shift will be equal to approximately 31.3 Hz.

The reason for the shift in frequency is that, as the target approaches, each successive wave has to travel a shorter distance to reach the target before being reflected and detected. The returned waves are essentially compressed in the direction of travel, and the distance between wave crests decreases, thus decreasing the wavelength and increasing the frequency.

If the radar is directed toward a target that is moving away, exactly the opposite situation occurs: each wave has to move farther, and the distance between wave crests increases, thus increasing the wavelength and decreasing the frequency. In either situation, the Gunnplexer generates the same beat tone for a certain velocity, whether the target is approaching or receding.

EXPERIMENTS AND QUESTIONS

1. Place an IF-508 diffraction mosaic slide at a distance of 1 m away from a sheet of white paper placed on the wall, as shown in Figure 4a. Using laser pointers of different wavelengths, measure the fringe separations of the interference pattern produced when the beam passes through the double-slit separations of $s = 4.5$, 5.8, 7.5, and 10 μm. How well do your measurements agree with $d = \lambda r/s$? How does the wavelength of the laser affect your ability to measure the fringe distance? Explain.

2. Replace the IF-508 in the prior setup by a human hair. Using laser pointers of different wavelengths, measure the fringe separations of the interference pattern produced when the beam passes through the hair. Calculate the thickness of the hair. Try thinner and thicker hairs. If available, try a spider web filament. How does the thickness of the hair affect the interference pattern? What combination of wavelength and hair thickness gives the widest fringe separation? Explain.

3. Set up the Gunnplexer units as shown in Figure 16. Measure the received signal, and then reposition the Gunnplexers to increase the distance between the units by 5 cm. Do this for distances between 20 and 60 cm (or more if your setup allows it). Do your data support the idea that the Gunnplexer's detector is sensitive to the intensity I of the electric

field, which falls in proportion to the inverse of the square of the distance $(1/r^2)$ between the transmitter and the receiver? Explain.

4. Place the receiver and transmitter on a goniometer stand so that they face each other at a distance of around 50 cm, with the pivot point exactly midway between them. Take a reading of the received signal strength, and then change the angle between the arms by 5°. Sweep the arm to obtain measurements between −90° and +90°. How tight is the microwave beam? How well does it match a Gaussian distribution? Explain.

5. Place the receiver and transmitter on a stand so that they face each other at a distance of around 50 cm. Take a reading of received signal strength, and then rotate the receiver to change its polarization angle by 5°. Continue to rotate the receiver to obtain measurements between −90° and +90°. How does relative polarization affect the intensity of the detected signal? Does the intensity as a function of polarization follow a smooth curve? Explain.

6. Set up the Gunnplexer units as shown in Figure 16. Measure the received signal when $r = 0$ and then *slowly* move the Gunnplexers away from one another. Record the distances at which the signal peaks and dips out to around 20 cm. Calculate wavelength by measuring the distance between successive peaks (or successive dips), which should equal $\lambda/2$. What is the frequency $f[Hz] = c/\lambda[m] = 3 \times 10^8 [m/s]/\lambda[m]$ at which your Gunnplexer is transmitting? How well does the calculation of frequency based on the average of your measurements compare to the specified frequency for your Gunnplexer? Explain.

7. Place the receiver and transmitter on a stand so that they face each other at a distance of around 50 cm. Rotate the receiver 90° away from the transmitter's axis of polarization. The signal at the receiver should drop close to zero. Insert a polarizer at 0°, 45°, and 90° referenced to the transmitter's polarization. How can inserting the polarizer at 45° increase the signal level at the detector?

8. Take three pieces of linear-polarizing film and look at the reflection of low-angle sunlight from a water surface with each one. Rotate each piece of film until reflections are minimized. Mark the vertical axis on each slide as its axis of polarization. Try the Figure 15 experiment with your labeled pieces of polarized film. Do your results agree with Figure 15? How does rotating one of the pieces of film in Figure 15a and Figure 15c change the amount of transmitted light? Explain.

9. On a clear day, look at different portions of the sky with a piece of polarizing film. Are any areas polarized? What is the polarization of light reflected from pavement? Explain your results.

10. Look at a pair of polarized sunglasses through one of your labeled pieces of polarizing film. What is the axis of polarization of the lenses in the sunglasses? Check out other sunglasses. Are they all polarized in the same orientation? Why do you think sunglasses are polarized in this way?

11. Polarizing film acts as a chemical version of the wire grid. Instead of long, thin wires it uses long, thin polyvinyl alcohol molecules that contain many iodine atoms. These long, straight molecules are aligned almost perfectly parallel to one another, with electrical conductivity provided by the iodine atoms. Examine a polarizing sheet very carefully. Can you see a pattern that resembles a wire grid? Would it be visible under a light microscope? Explain.

12. Place the receiver and transmitter on a goniometer stand so that they face each other at a distance of around 50 cm, with the pivot point exactly midway between the receiver and transmitter. Place a double-slit slide with $s = 75$ mm over the pivot point, as shown in Figure 18. Take a reading of received signal strength at $\theta = 0°$, and then change the angle between the arms by $5°$. Sweep the arm to obtain measurements between $-90°$ and $+90°$. Repeat, using a double-slit slide with $s = 105$ mm. Plot your data, and identify the angles at which the peaks and dips of the interference pattern occur. Are successive peaks (or successive dips) separated according to $\sin \theta = \lambda/s$? Explain.

13. Connect the Gunnplexer's detector diode output (mixer output) to an audio amplifier. Point the Gunnplexer at a passing car and pay attention to the whooshing sound that you will hear. How does the Doppler sound produced by the Gunnplexer from an approaching car compare to the sound produced when the Gunnplexer is pointed at a receding car? Explain.

CHAPTER 2

LIGHT AS PARTICLES

Next time you turn on an electric oven, pay attention to the way in which the color of its glow changes as it heats up. The glow is at first weak and dull red before it turns bright orange. At higher temperatures, such as those in a blacksmith's metalworking oven, the glow becomes much brighter and whiter.

All materials emit electromagnetic radiation with a very similarly colored glow as a function of temperature. To simplify the study of this phenomenon, physicists make use of an idealized object known as a *blackbody*. This theoretical device is able to absorb all electromagnetic radiation that reaches it, and then re-emits this radiation in a characteristic, continuous spectrum. Because this object reflects no electromagnetic radiation (including visible light) it theoretically appears black when it is cold, and this is where it gets its name. However, at any temperature above absolute zero, the blackbody emits a temperature-dependent radiation known as *blackbody radiation.*

In 1905, British Nobel laureate Lord Rayleigh (his real name was John William Strutt) and his colleague Sir James Jeans derived an expression for blackbody radiation. They assumed that blackbodies are made of a huge number of *harmonic oscillators*—imaginary contraptions somewhat like guitar strings that can vibrate in response to a sound and can themselves produce a sound when they vibrate.

As shown in Figure 20a, a harmonic oscillator—such as an ideal elastic string held between two walls (which force zero movement at the ends)—is able to vibrate at any wavelength where the string's ends become zeros of the wave. The frequency of the vibration with the longest wavelength is called the *fundamental*, and its integer multiples are called *harmonics* or *overtones*.

The classical physics* used by Rayleigh and Jeans requires all harmonic oscillations from a system at equilibrium (one where the energy being gained by the system equals the energy lost by the system) to have an average that is proportional to the system's temperature. Since there are an infinite number of possible harmonics, this means that the energy of each harmonic must be the same.

Unfortunately, this requirement of classical physics also means that most of the harmonic oscillators will be buzzing along at the smaller wavelengths (which correspond to higher frequencies), where most of the infinite harmonics exist. As a

---

*Specifically, a branch called "statistical mechanics," which deals with the behavior of groups of many particles. For example, the pressure of a gas is not caused by the behavior of a single molecule of the gas, but rather by the statistical behavior of a huge number of molecules at a specific temperature.

Exploring Quantum Physics Through Hands-On Projects. By David Prutchi and Shanni R. Prutchi
© 2012 John Wiley & Sons, Inc. Published 2012 by John Wiley & Sons, Inc.

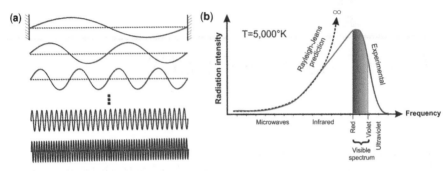

Figure 20   In 1905, Rayleigh and Jeans derived an expression for the blackbody radiation. (a) They assumed that the blackbody is composed of a huge number of electromagnetic harmonic oscillators—somewhat similar to guitar strings—that can support vibrations of any wavelength that fits completely between their two ends. (b) Although the Rayleigh–Jeans expression matches experimental data at low frequencies, it incorrectly predicts infinitely large amounts of radiation at high frequencies.

consequence of this, the Rayleigh–Jeans expression predicts an energy output from the blackbody that rises toward infinity as the wavelength approaches zero (Figure 20b).

Since the whole universe is not scorched to a crisp every time we light the oven to bake cookies, something must be terribly amiss with the assumptions made by classical physics regarding the behavior of blackbodies! Einstein, Rayleigh, and Jeans understood this to be a problem that required a breakthrough idea. The failure of early twentieth-century physics to explain blackbody radiation became known as the "Ultraviolet Catastrophe," since it is obvious that the major discrepancy between theory and experiment occurs at the short wavelengths of the UV.

Let's see how a blackbody actually radiates as a function of temperature. For this experiment, the filament inside a flashlight bulb will serve as a blackbody that can be safely heated in a controlled manner up to quite high temperatures.

Our experimental setup is shown in Figure 21. We constructed a goniometer stand by cutting in half a wooden metric ruler (these are sold by educational supply stores). We then attached the two halves to a steel protractor, and reinforced the pivot axis with duct tape. We used low-cost optical component holders from a student optical bench sold by Anchor Optics as model AX26095.

A 14-V 200-mA #1487 screw-base flashlight lamp (RadioShack 272-1134) sits on a mating base (RadioShack 272-356) within a 2-in. schedule 40 PVC pipe cap as shown in Figure 21b. A small hole through the cap (e.g., 3/32-in. diameter) provides a direct view to the lamp's filament. The lamp is powered by a low-voltage, variable, and metered power supply. Light from the lamp assembly is projected by a double-convex lens onto a 30-mm glass equilateral prism (Anchor Optics AX27689), which disperses the light into its constituent colors. The spectrum is focused by a second double-convex lens to form a sharp "rainbow" line at the input slit of the light detector.

We measure light intensity with the same integrated light sensor as we did in chapter 1 for the interference-pattern scanner (Figure 5). The device we use is a

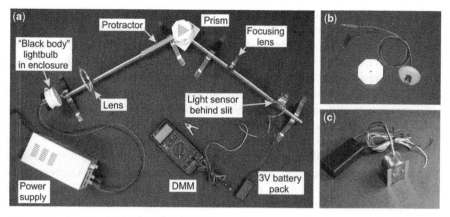

Figure 21   Studying the radiation produced by a blackbody is possible using a simple goniometer setup that uses a glass prism to separate light from the blackbody into its component color spectrum (**a**). (**b**) The filament of a flashlight lightbulb acts as the blackbody. (**c**) A low-cost, integrated light-to-voltage converter is used to measure the light intensity at various wavelengths.

TAOS TSL254R-LF light-to-voltage converter. This device is an inexpensive component that incorporates a light-sensitive diode and amplifier on a single chip. It is very easy to use. It requires a supply voltage in the range of 2.7 to 5.5 V (we use two 1.5-V AA batteries in series), and produces an output voltage that is directly proportional to the light intensity (Figure 22a). As shown in Figure 21c, we placed the sensor's lens behind a small hole that we drilled on a $1\frac{1}{4}$-in. copper pipe cap. We

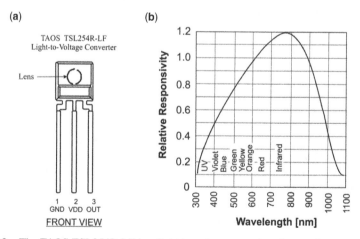

Figure 22   The TAOS TSL254R-LF is a light-to-voltage converter that produces an output voltage directly proportional to the light intensity entering through its integrated lens. (**a**) Power in the range of 2.7–5.5 V is supplied between pin 2 and ground. (**b**) The output at pin 3 is a voltage that is proportional to light intensity at a rate of 9 mV/(μW/cm^2) at $\lambda = 635$ nm. The responsivity must be corrected based on this graph for other wavelengths.

also drilled and tapped two small holes on the sides of the center hole to accept two 2-56 machine screws. These hold two single-edge razor blades that form a slit allowing only a narrow portion of the light spectrum to enter the sensor. We closed the other end of the copper cap with copper tape to ensure that light enters the assembly only through the slit between the two razor blades. We found that adding a bit of matte-finish transparent adhesive tape (Scotch® Matte Finish Magic™ Tape) over the slit improves measurements by diffusing the light before it reaches the sensor's lens.

None of the components are critical, so you may adapt the design in any way you see fit. The characteristic change in color of the light emitted by a blackbody can be seen by simply varying the supply voltage to the lightbulb. At some point, the filament is hot enough to produce a soft, dull red glow. As the supply voltage is increased, the filament glows brighter, with a more red-orange hue. At even higher voltage, the filament produces intense white light.

The filament is made of tungsten—a metal that has the highest melting point of all the nonalloyed metals. The electrical resistance of tungsten varies with temperature, so we can calculate the filament's temperature T by monitoring the power supply's voltage V and current I to measure its resistance when hot:

$$T = 300^\circ \text{ K} + \frac{\frac{\left(\frac{V}{I}\right)}{R_0} - 1}{4.5 \times 10^{-3}} [^\circ \text{K}]$$

where R_0 is the resistance of the filament that you'll need to measure very carefully with a multimeter at room temperature (around 300° K).

Place a white sheet of paper at the distant end of the goniometer arm opposite where you will place the light sensor. With the bulb shining brightly, adjust the prism, goniometer angle, and location of the lenses to produce a crisp light spectrum. Dim the room lights and vary the lightbulb voltage. Do you see how the short-frequency wavelengths (green, blue, violet) disappear as the voltage is lowered?

Write down the angle of the protractor at which each color falls on the sensor's slit. The sensor's range is wider than the spectrum visible to the human eye, so you can take an IR measurement below the lowest red you can see, as well as a UV measurement beyond the deepest visible violet. Equip yourself with a small flashlight and turn off the room lights. Read the light intensity for every color (be consistent with the angle between measurements) at a few different filament voltages.

The voltage output of the TSL254R-LF varies with wavelength, so before you plot your data, correct it according to the sensor responsivity curve of Figure 22b. For example, the measurement for the intensity of light at 850 nm (IR) should be divided by 1.2 to represent the same voltage output per corresponding light intensity as a measurement made at 600 nm (orange). Our results are shown in Figure 23, and show the characteristic distribution that puzzled early twentieth-century physicists.

In 1893, German physicist Wilhelm Wien noted that the wavelength distribution of the radiation from a blackbody at any temperature has essentially the same shape as the distribution at any other temperature, except that each wavelength moves over on the graph. Wien demonstrated that the peak of the spectrum of the light emitted

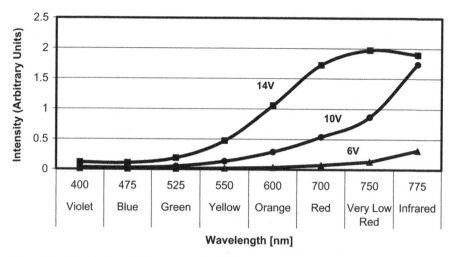

Figure 23   Blackbody radiation measured with the setup of Figure 21 at three filament voltages.

by a blackbody occurs at a wavelength:

$$\lambda_{\text{MAX}} = \frac{b}{T} = \frac{2.898 \times 10^{-3} \, [\text{m}^\circ \text{K}]}{T[^\circ\text{K}]}$$

where T is the absolute temperature of the blackbody, and b is Wien's displacement constant, equal to 2.898×10^{-3} m° K. Wien also came up with an empirical formula for the blackbody spectrum. Although his equation does follow quite closely the shape of the spectrum at short wavelengths, it fails to accurately fit the experimental data for long-wavelength emissions.

THE SEED OF QUANTUM PHYSICS: PLANCK'S FORMULA

Returning to our story about the Ultraviolet Catastrophe, Einstein suggested solving it by using a mathematical "trick" that had been proposed 5 years earlier by Max Planck when he was trying to improve upon the Wien approximation. Planck produced a formula that matched the data very well by assuming that the harmonic oscillators could not oscillate at any frequency, but instead were limited to a set of discrete, integer multiples of a fundamental unit of energy, E, proportional to the oscillation frequency f:[†]

$$E = hf$$

[†]Frequency is often represented by the Greek letter "nu" (υ). However, to keep things consistent, we'll represent frequency by the Latin letter "f."

where $h = 6.626 \times 10^{-34}$ J·s, is the constant that is now known as *Planck's Constant*. Planck introduced this as a mathematical ploy to reduce the number of possible harmonic oscillators vibrating at high frequencies in the blackbody, thus reducing the average energy at those frequencies by application of the equipartition theorem. In this way Planck was able to come up with a formula where the radiated power decreases toward zero at high frequencies, yielding a finite total power that closely matches experimental data. Planck's Radiation Law predicts the intensity I radiated by the blackbody at each wavelength λ for any temperature T according to:

$$I(\lambda, T) = \frac{2\pi hc^2}{\lambda^5} \cdot \frac{1}{\left(e^{\frac{hc}{\lambda kT}} - 1\right)}$$

where $k = 1.38 \times 10^{-23}$ J/°K is Boltzmann's constant.

Planck believed that this *quantization* (allowing only discrete, integer multiples of frequency) applied only to the imaginary harmonic oscillators that were thought to exist in the walls of the blackbody cavity. Planck did not attribute any physical significance to this assumption, and believed that it was merely a mathematical device that enabled him to derive a single expression for the blackbody spectrum that matched the empirical data at all wavelengths.

As we will see in chapter 4, we now know that Planck's assumed quantized oscillators are the very atoms that make the blackbody, and that these atoms radiate energy when electrons jump between a higher energy level to a lower energy level.

Go back over this short section, and try to understand it really well! The simple expression $E = hf$ is at the very core of quantum physics! You will see this expression, as well as Planck's constant $h = 6.626 \times 10^{-34}$ J·s very often in the sections and chapters that follow.

THE PHOTOELECTRIC EFFECT

In 1905, Albert Einstein took Planck's proposal one step further by stating that light itself is a beam of discrete packets, like bullets from a machine gun. Einstein needed this assumption to resolve another troublesome inconsistency in physics, this one concerning the *photoelectric effect*, by which electrons may be ejected from a metal surface illuminated by light.

Heinrich Hertz observed in 1887 that the minimum voltage required to draw sparks from a pair of metallic electrodes was reduced when they were illuminated by UV light from a mercury vapor lamp (Figure 24a). You may remember that in the same year Hertz confirmed the electromagnetic-wave nature of light. As such, it was no surprise to him that the energy carried by a light wave could shake electrons loose from an electrically-charged piece of metal.

Take a look at Figure 24b and imagine the following situation: you are on a secret mission in the Amazon jungle. Your arch-enemy shows up as you perilously make your way through a rope bridge over piranha-infested waters. He starts to shake the bridge as you hold on for dear life. After some effort on your nemesis'

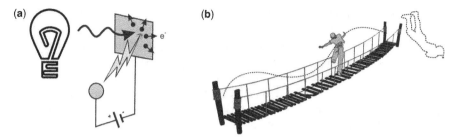

Figure 24   The photoelectric effect was first reported by Hertz. **(a)** He noticed that sparks could be drawn from an electrode at a relatively low voltage when illuminated by UV light from a mercury lamp. **(b)** Similar to a rope bridge, a wave of sufficient amplitude was thought to be able to shake electrons loose from the metallic surface.

part, the bridge will oscillate with such high amplitude that you will plunge to the cliff below. Fortunately, you remembered to pack a parachute and a bottle of piranha repellant . . .

Just as in this analogy, the wave theory of light would predict that the intensity of the light should determine if the light can generate these photoelectrons. This is because in the wave theory of light, intense light has more energy, and energy is what is needed to liberate the electrons from the metallic surface. Conversely, if the light is dimmed, a point should be reached at which the energy of the wave is too low to kick any electrons away from the metallic electrode.

Also like in the bridge analogy, the wave would take its time to build up sufficient strength to counteract the grip of the metal on its electrons. It would thus take some time after the electrode is exposed to light for it to start expelling electrons. After all, the wave energy is spread over a wide area, so it should take a long time for any one electron to be ejected.

Lastly, ignoring resonances and similar effects, it's not the frequency at which the bridge oscillates, but rather its amplitude that would cause you to lose your grip and fall. In the same way, the classical wave theory of light would predict that the emission of the electrons is dependent on the intensity of the light, not on its color.

Counter to classical wave theory however, experiment had shown that the intensity did not matter to the ability of light to release electrons from metal. All that mattered was the wavelength of the light. Long-wavelength light could not generate photoelectrons, even if the light was very intense. On the other hand, short-wavelength light could produce photoelectrons, even if the light was of very low intensity. Moreover, electrons are ejected as soon as the light shines on the metal.

In other words, if the light is too red, no electrons are ejected, no matter how bright the light. If the light is blue enough, electrons are ejected as soon as the light shines on the metal, and their energy increases as the color of the light moves toward the UV.

It took the genius of Einstein to understand that the experimental observations could be explained if light is a stream of particles, each with the energy given by Planck's formula $E = hf$. In this case, a light particle with sufficient energy would be able to knock out an electron from the metal. Einstein thus proposed that each particle of light carries a *quantum* of energy according to the color of its light. He named the light particles *quanta*. We now call them *photons*.

Putting it in terms of light that we can see, red light is composed of individual particles (also called quanta, or photons), where each individual particle (quantum, or photon) has less energy than a particle of violet light.

According to Einstein, red light is composed of low-energy quanta, so no one quantum would be energetic enough to knock electrons out of the electrode from Figure 24a. Increasing the intensity of the light does nothing more than increase the number of light quanta showering on this electrode, all of them too weak in energy to liberate an electron from the metal.

Beyond a certain threshold frequency, the individual light quantum is energetic enough on its own to liberate a single photoelectron. Above this frequency threshold, a change in intensity would change the number of photoelectrons that are released. However, since only one light quantum of sufficient energy is needed to liberate one photoelectron, the photoelectric emission with high-frequency light will happen no matter how weak the intensity of the light.

In 1921, Einstein won the Nobel Prize for his discovery of the law of the photoelectric effect.

From now on, to keep things consistent, we'll call particles of light by their modern name: *photons*. This name applies whether we talk about a photon corresponding to the very low frequencies of radio and microwaves, the visible spectrum, or the extremely high frequencies of cosmic gamma rays.

You can experiment with the photoelectric effect using a *phototube*, such as that shown in Figure 25. The device is a gas-filled or vacuum tube that has an electrode (the photocathode) which expels electrons when illuminated by a light source. The photoelectrons can be collected by a thin wire electrode (the anode), causing a current to flow through a closed circuit. The cathode is typically coated with one or more alkali metals (e.g., lithium, potassium, and cesium). Unlike other metals, which

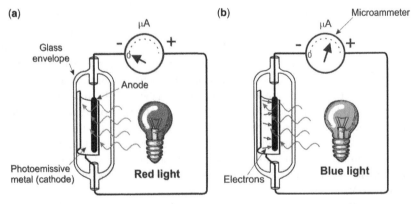

Figure 25   The photoelectric effect requires light to act as a stream of particles. **(a)** Low-frequency photons (red light) do not supply enough energy to eject electrons from the metal cathode of a certain phototube. **(b)** Blue photons have enough energy to overcome the "work function" of the metal and eject photoelectrons.

emit photoelectrons only when illuminated by UV light or X-rays, the energy of visible-light photons suffices to release electrons from alkali metals via the photoelectric effect.

For an electron to be ejected from the photocathode, the energy it gains must be greater than some energy W_o necessary for the electron to free itself from the metal. W_o is called the *work function* and is a property of the metal from which the electrons are to be ejected. Since the energy of a photon is given by $E = hf$, photoelectrons can be liberated only if $E > W_o$. A photon with energy equal to W_o is barely able to liberate an electron from the metal. However, if the photon carries more energy than W_o, the electron will be ejected with some kinetic energy.

For example, if the metal plate is made of sodium, which has a work function $W_o = 3.296 \times 10^{-19}$ J, the threshold light frequency for photoelectric emission would be:

$$f_{\text{threshold}} = \frac{W_o}{h} = \frac{3.296 \times 10^{-19}[\text{J}]}{6.626 \times 10^{-34}[\text{J} \cdot \text{s}]} = 497.4 \times 10^{12}\text{Hz}$$

which corresponds to a bright red/orange color. As such, a deep red light will not cause photoelectric emission, showing no current in the microammeter of Figure 25, while blue light will easily cause a photocurrent to flow through the circuit.

Photoelectric tubes were available before the days of semiconductor photodetectors. They were commonly used to read the optical sound tracks of film, and as sensors in burglar alarms and door openers. They are still manufactured for some specific applications (e.g., UV flame detection, spectrophotometers, etc.), but the easiest and cheapest source for these units is the surplus market, for example from Sphere Research in Canada or on eBay. The photocathodes in common phototubes are made with a mixture of alkali metals to make them sensitive over a wide range of wavelengths, usually from IR to UV. To be able to conduct the experiment as shown with red and blue light, you will need to find a phototube with a so-called "S-12" response.

A practical way of experimenting with any available phototube is to place it in a setup similar to that shown in Figure 26. Here, a commonly available RCA 1P39 phototube is illuminated by an LED[5]. A few inexpensive LEDs of different colors[‡] will allow you to collect photoelectric data with photons at different wavelengths. The LED should be positioned so that most of its light falls on the photocathode. Stick a thin (3-mm) strip of black electrician's tape to the glass envelope of the phototube to shield the anode electrode from direct light from the LED. This reduces the likelihood that spurious photoelectrons from the anode affect your measurements. In addition, you should paint the inside of your enclosure with flat black paint or cover it with matte black tape to prevent reflections from entering the tube.

The idea is to use the famous equation developed by Einstein to describe the photoelectric effect:

$$eV_s = hf - W_o$$

[‡]Make sure the LEDs are packaged in clear, not colored, plastic.

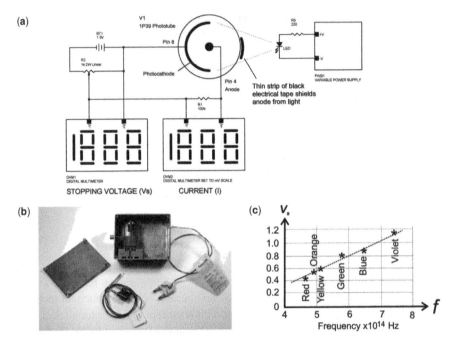

Figure 26 Build this setup to experiment with the photoelectric effect. (a) A commonly available 1P39 phototube is shown, but many others will work equally well. (b) The phototube and LED should be placed in a light-tight box that is painted flat black on the inside. A thin strip of black electrician's tape should be used to shadow the phototube's anode electrode. (c) According to Einstein's equation for the photoelectric effect, all V_s for various LED frequencies f should fall on a line with slope h/e.

where $e = 1.60 \times 10^{-19}$ Coulomb is the charge of the electron, V_s is the negative voltage that needs to be applied to the anode to stop current from flowing through the circuit ($I = 0$), W_o is the work function we discussed above (the minimum energy needed to liberate the electron from the photocathode),[§] and hf is the energy of the photons striking the photocathode.

By the way, the kinetic energy of an electron accelerated by a voltage V is eV, so Einstein's equation tells us that the energy necessary to stop the most energetic photoelectron (eV_s) equals the energy of the photon that kicks it out of the metal (hf), minus the energy required to liberate the electron from the metal (W_o).

Einstein's equation predicts that plotting the stopping potential V_s as a function of frequency f should result in a line with slope h/e. If you can't find the LED's specs for the exact wavelength of the light emitted, use the values shown in Table 2. Try it

---

[§]Our experimental setup does not really measure W_o of the photocathode, but rather a closely related value that includes the work function of the electrode collecting the photoelectrons. This is ignored by many textbooks, but is an important distinction that must be made when dealing with real-world applications of the photoelectric effect.[6]

TABLE 2 **Wavelengths and Frequencies for the Light Emitted by Some Common LEDs**

LED color	Wavelength λ	Frequency $f = \dfrac{c}{\lambda} = \left(\dfrac{2.998 \times 10^8 \left[\frac{m}{s}\right]}{\lambda[m]} \right)$
Violet	405 nm	7.40×10^{14} Hz
Blue	470 nm	6.38×10^{14} Hz
Green	563 nm	5.33×10^{14} Hz
Yellow	585 nm	5.13×10^{14} Hz
Orange	620 nm	4.84×10^{14} Hz
Red	650 nm	4.62×10^{14} Hz
IR	945 nm	3.18×10^{14} Hz

out with your apparatus and see how well your data match the equation that won Einstein his Nobel Prize![¶]

At what light frequency would you expect V_s to drop down to zero if you extend the line you obtained when graphing your data? This would be the threshold frequency for photoemission for your phototube. You can prove the existence of a threshold by using your IR LED (945 nm). Since the emission from the IR LED is invisible to the human eye you will need to use a digital or video camera (especially one with "night-vision mode") to verify that your LED is shining bright IR light on the phototube. Close the box and take a reading of photocurrent I with a $V_s = 0$. Nothing! Nada! Zero! Zip! The LED's photons at $f = 3.18 \times 10^{14}$ Hz are not energetic enough to kick electrons free from the photocathode.

Another interesting experiment that you should try is to find if varying the intensity of the light has an effect on V_s. Keep the frequency f of the photons constant by using only the green LED. Reduce the intensity of the light by limiting the LED supply current. What do you think will happen?

Lastly, remember that the wave theory predicted that it would take some time for the photocathode to be exposed to light before it starts expelling electrons? Well, you can prove this prediction wrong too by substituting the power supply in the setup of Figure 26 with a fast pulser, and by monitoring the photocurrent with an oscilloscope. This is possible, because unlike incandescent lightbulbs that need to heat up before their light peaks, LEDs turn on as soon as current is applied, and their emission turns off as soon as current is interrupted.

Turning on an LED via a switching device (e.g., a transistor) using its nominal voltage (2.1 V in the case of a green LED) would cause the LED to start emitting photons only after a few microseconds. This is because all of the *parasitic capacitances*— unintended capacitors that form between the various elements in real-world LEDs and the power supply circuit—need to be charged to the LED's threshold voltage before light is emitted. To overcome this problem, the circuit of Figure 27a dumps a

---

[¶]No, it was not $E = mc^2$. The 1921 Nobel Prize in Physics was awarded to Albert Einstein "for his services to Theoretical Physics, and especially for his discovery of the law of the Photoelectric Effect."

(a)

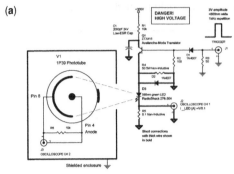

(b)

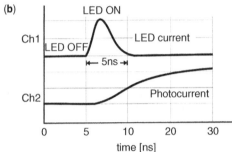

(c)

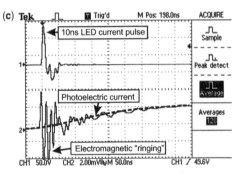

Figure 27   You can measure the time it takes a photoelectric tube to generate a current in response to light. **(a)** A green LED is turned on very quickly by passing a very brief, very strong pulse of current. At the same time, the photoelectric response is observed with an oscilloscope. **(b)** Although the response time is greatly slowed down by real-world effects, a nanosecond-level response time is still many orders of magnitude faster than predicted by classical wave theory. **(c)** Pulsing an LED with a very brief, very strong current causes the circuit to radiate an electromagnetic pulse that couples into the low-level signal line from the phototube, causing a "ringing" artifact that obscures the signal.

2,000-pF capacitor charged to 300 V into the LED, thus causing the LED to produce a pulse of light that starts within 1 ns after the current's leading edge. The LED will continue to produce light for around 5 μs, as these parasitic capacitances keep on supplying current until they are discharged.

When you build the circuit, be careful to keep the wires shown in bold very short. In addition, use thick wires for these connections. Please note that the transistor is not a plain NPN transistor. It is a device designed specifically to work in the "avalanche" mode, which makes it possible to produce very brief, high-power pulses. You must supply the circuit with a minimum voltage of 270 V, which is the avalanche breakdown voltage for the ZTX415 transistor. The trigger pulse is just a >3-V pulse that can be delivered at repetition rates of 100 to 1,000 pulses/min.

As shown in Figure 27c, pulsing an LED with a very brief, very strong current causes the circuit to radiate an electromagnetic pulse that couples into the low-level signal line from the phototube, causing a "ringing" artifact that obscures the signal. For this reason, let's concentrate instead on the simplified results of Figure 27b, and try to understand what they show. The photocurrent kicks in some 1 ns after current is applied to the LED. How does this 1 ns compare to the predictions of classical wave theory? Go back to the rope bridge analogy of Figure 24b. A wimpy villain would not be able to build enough wave amplitude on that heavy bridge to make you fall before you cross the bridge. He would need as much energy as possible to build up wave amplitude quickly, so he will probably have some of his brutish minions shake the bridge. How quickly you fall off the bridge depends on how fast the brutes are able to transfer energy to the ropes. Energy transfer over a certain period of time is called power. Evil Dr. Wave must surely hope his cronies ate their Wheaties® for breakfast!

Let's use this analogy to calculate the time it would take a wave to build up enough amplitude in our phototube: let's suppose that all of the energy in the 2,000 pF capacitor is converted into light by the LED. This is not a realistic assumption due to the inherent losses of the circuit and the LED's intrinsically low conversion efficiency, but let's suppose 100% efficiency nevertheless. The optical pulse's power is thus:

$$\frac{\text{Energy in capacitor}}{\text{Time LED remains on}} = \frac{\left(\frac{1}{2}CV^2\right)}{5\,\mu s} = \frac{\left(\frac{1}{2} \times 2{,}000 \times 10^{-12} \times 300^2\right)[\text{J}]}{5 \times 10^{-6}[\text{s}]} = 18\text{W}$$

Let's also assume ideal conditions, and suppose that all of this light could be transferred to build up enough of a wave amplitude to shake electrons from the atoms on the surface of the photocathode. Let's assume that the atoms in the photocathode are evenly distributed and each takes $1 \times 10^{-19}\,\text{m}^2$ of the photocathode's surface area. If the light from the LED illuminates $3\,\text{cm}^2$ of photocathode surface, then each atom can absorb energy at a maximum rate of:

$$\text{Total power} \times \frac{\text{Area taken by atom}}{\text{Area of photocathode}} = (18\,\text{W})\frac{1 \times 10^{-19}\,\text{m}^2}{3 \times 10^{-4}\,\text{m}^2} = 6 \times 10^{-15}\,\text{W}$$
$$= 6 \times 10^{-15}[\text{J/s}]$$

The work function W_o of the photocathode in our tube is 2.24×10^{-19} J, so according to classical wave theory, it would take an electron being supplied at 6×10^{-15} W at least 2.24×10^{-19} J$/6 \times 10^{-15}$ [J/s] $= 37.333\,\mu s$ to escape from the metal under absolutely optimal conditions.

These $37.333\,\mu s$ are 37,333 times longer than the 1 ns that our real-world experiment showed as a worst-case response time! This factor increases by a few orders of magnitude if you account for the LED's inefficiencies, turn-on time, and so on. Once again, the wave theory of light is discredited!

CAN WE DETECT INDIVIDUAL PHOTONS?

Let's do a back-of-the-envelope calculation of just how many photons are produced by a typical 60-W lightbulb: although we know that the filament acts as a blackbody, and thus produces a wide range of wavelengths, let's take 600 nm as an average for our quick estimate. The energy of a photon is $E = hf = hc/\lambda$, so the energy (in Joules) of a single photon of 600-nm light is $E_{\lambda = 600\text{nm}} = 6.626 \times 10^{-34}$ [J · s] $\times 2.998 \times 10^8$ [m/s]$/600 \times 10^{-9}$ m $= 3.31 \times 10^{-19}$ J.

A power of 60 W means 60 J/s, so the number of photons produced by a 60-W lightbulb is approximately 60 [J/s]$/3.31 \times 10^{-19}$ [J/photon] $= 1.81 \times 10^{20}$ photons per second. That is a huge number!

Only a small fraction of these reach our eyes, even if we look directly at the lightbulb. This is because the photons are spread over a sphere, and only a small fraction of the photons reach our eyes at a safe distance from the filament. Imagine that you are driving around at night, and you want to see if your friend is still awake. You look at the window from a block away (let's assume 100 m) and see that the light is still on. How many photons reach your eye?

The photons can be assumed to be flying away from the lightbulb equally in all directions. Since you stand 100 m away, let's place the lightbulb at the center of a sphere with a radius $r = 100$ m. The sphere has a surface area of $4\pi r^2$, so an $r = 100$ m sphere has a surface area of 125,664 m^2. The pupil of the human eye has an effective diameter of 8 mm ($r = 4$ mm) in the dark, so its area πr^2 is around 50×10^{-6} m^2. We need to scale down the total number of photons emitted by the lightbulb by 50×10^{-6} m$^2/125{,}664$ m$^2 = 400 \times 10^{-12}$. That means that $400 \times 10^{-12} \times 1.81 \times 10^{20} = 72.4 \times 10^9$ photons from a 60-W lightbulb at a distance of 100 m reach your eye every second. That's still a lot of photons from a relatively dim source!

A typical phototube barely gives a measurable current at that level, so how could we ever expect to detect a single photon? Fortunately, some very smart engineers at Westinghouse and RCA figured out that the single electron released from a photocathode by a single photon could be accelerated toward another electrode in order to produce secondary electrons. Two or more electrons are then released when the accelerated photoelectron slams into the electrode. As shown in Figure 28, the same process can be repeated over and over again with the secondary electrons used to successively multiply the number of electrons released in a cascade. A much larger number of electrons finally reach the anode as the result of a single photon hitting the photocathode. Commercially available *photomultiplier tubes* (PMTs) based on this principle produce as many as 10^6 to 10^7 secondary electrons at the anode for each photon that releases a photoelectron from the photocathode.

Today, the PMT is the workhorse detector in experimental particle physics. One of the most extreme is the Super-Kamiokande detector used to hunt for elusive particles called *neutrinos*. This massive detector is buried deep within an abandoned zinc mine in Japan. It uses 11,242 PMTs to look at photons produced when the neutrinos decelerate as they hit 50,000 tons of pure water.

Let's build the PMT probe (Figure 29) that we'll use for many experiments in the following chapters. We chose the RCA 6655A PMT because it is affordable

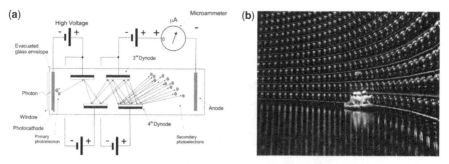

Figure 28   Photomultiplier tubes are the workhorse detector of experimental particle physics. **(a)** The photoelectrons released at the photocathode of a PMT are accelerated toward an electrode (first dynode), and cause the release of two or more secondary electrons. Each of these causes the release of two or more electrons from the second dynode. The cascade continues until a very large number of electrons are available for detection at the anode. **(b)** Technicians on a rubber boat inspect some of the 11,242 PMTs in the Super-Kamiokande neutrino detector in Japan.

and widely available in the surplus market. This tube has ten dynodes that multiply a single photoelectron into 1.6×10^6 secondary electrons at the anode when operated at $+1,000$ V. You can increase the voltage beyond $+1,000$ V, and operate the PMT safely at $+1,250$ V, but be careful not to exceed its absolute maximum voltage of $+1,600$ V.

   As shown in Figure 30, we enclosed our PMT probe in a die-cast aluminum box (Bud Industries model AN-1323). The PMT and its magnetic shield fit snugly within a 10-cm-long piece of aluminum optical instrument construction rail (66-mm profile,

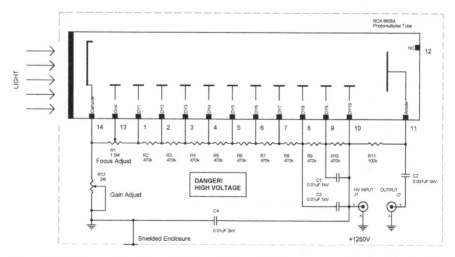

Figure 29   This versatile photomultiplier probe is useful for many experiments described in this book. It is based on the RCA 6655A PMT, and features a gain of 1.6 million when operated at $+1,000$ V.

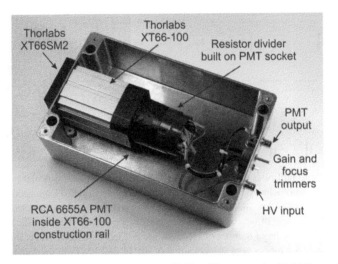

Figure 30   Inside view of our PMT probe. The PMT and its magnetic shield fit snugly within a 10-cm-long piece of 66-mm aluminum optical instrument construction rail. A matching faceplate is used as an interface to other optical components. All possible light leakage paths around the PMT are sealed with black electrician's tape. The resistive voltage divider and filter capacitors are assembled directly on the PMT's tube socket.

Thorlabs model XT66-100). A matching faceplate (Thorlabs model XT66SM2) screws to the 66-mm-profile rail from outside the aluminum box. It is used as a convenient, light-tight interface to other optical components. The plate is sealed against the metallic enclosure with black RTV silicone caulking, which can be purchased at any auto supply store. All other possible light leakage paths around the PMT are sealed with black electrician's tape. There is enough room inside the box to accommodate the PMT's tube socket, resistor divider, capacitors, potentiometers, and coaxial connectors. The bottom cover for the enclosure is not shown in the picture, but we drilled a blind hole in the center and tapped it to accept a $\frac{1}{4}''$ 20 TPI machine screw, so that we may mount the probe on a standard camera tripod or on an optical table post.

A word of caution: never expose the PMT's sensitive face to bright light! This will cause permanent damage to the tube, especially when the tube is powered. While not in use, our PMT is protected with a faceplate cap with an SM-2 series end-cap (Thorlabs model SM2CP2).

LOW-COST PMT POWER SUPPLIES

Ideally, you should power your PMT from a lab-grade, low-noise, high-voltage power supply designed specifically to bias detectors. However, achieving exceptionally low ripple and high stability in a high-voltage power supply is not trivial, making these power supplies pricey. Designs for these low-noise power supplies are secrets closely guarded by companies that specialize in such products, such as Matsusada Precision,

EMCO High Voltage, and Hamamatsu Photonics. Fortunately, relatively low-cost, high-voltage power supply modules are sold by these companies as components for larger instruments.

Our home-built PMT power supply is built around the Matsusada JBE-2P high-voltage power supply module. This module sells new for around $200, but similar units can be found on the surplus market for a fraction of this price. The module is powered by 24 VDC (at 300 mA typical), and generates up to 2,000 V at 1 mA as a function of a 0- to 10-V control input. A 10-V reference output is available on the JBE-2P, allowing a 5 kΩ potentiometer to be used for voltage control.

The schematic diagram of Figure 31 shows that 24 V for the JBE-2P are obtained by rectifying the output of a 25.2-V transformer (T1, a RadioShack 273-1366A) and regulating through a LM317 variable linear regulator. A kit comprising a circuit board and components for the regulator circuit is conveniently available from Velleman as their model K1823.

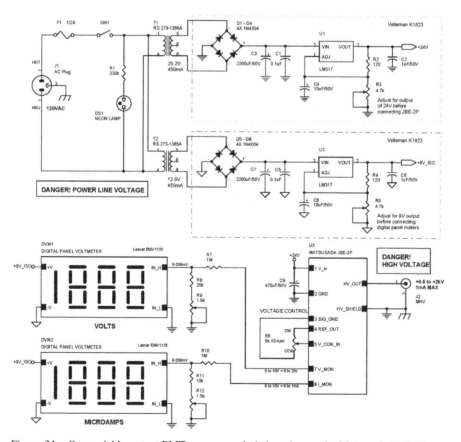

Figure 31   Our variable-output PMT power supply is based around a Matsusada JBE-2P low-ripple, high-stability, high-voltage power module. Output voltage (up to 2 kV) and current (up to 1 mA) are monitored via two LCD panel meters.

A second transformer and regulator board are used to produce an isolated output of +9 VDC to power two $3\frac{1}{2}$-digit panel voltmeters. The digital voltmeters are used to show the voltage and current output from the high-voltage module. We used Lascar EMV1125 LCD panel meters because they need only a 7/32-in. round hole for mounting, but any other 200-mV full-scale panel meter module should work equally well.

An even lower-cost alternative is to build the PMT power supply from a BXA-12579 high-voltage inverter used to drive a cold-cathode fluorescent lamp (CCFL). This under-$20 module produces 1,500 VAC at around 30 kHz from a 12-VDC input.

As shown in the schematic diagram of Figure 32, the BXA-12579 has to be slightly modified for use in this circuit. The 100-μF/25-V capacitor marked "C1" in the BXA-12579 needs to be removed so that the power input to the module can be varied rapidly. In addition, a pin needs to be added to access the transformer's high-voltage terminal, and the line connecting the low-voltage to the high-voltage ground must be cut.

In this circuit, the supply to the BXA-12579 inverter is modulated by op-amp U3. The high-voltage AC output of the inverter is rectified and doubled by D2/D3 and C7/C12. C13/C14 are used to filter the high-voltage output.

The output voltage is regulated to the desired level by feeding op-amp U3 with an attenuated version of the output. This level is compared against the output of voltage control R17. The op-amp attempts to maintain the output constant by controlling the power to the inverter by way of transistor Q1.

This design incorporates a circuit used by Spectrum Techniques to reduce ripple in their PMT power supplies.[7] Op-amp U2 is used to neutralize AC signals from the

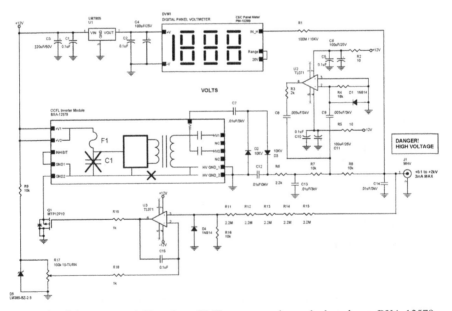

Figure 32 A low-cost, variable-voltage PMT power supply can be based on a BXA-12579 inverter module that was originally designed as a power supply for CCFLs.

high-voltage output by adding a voltage that is $180°$ out of phase with any signal that goes through C9. A very stable and clean output can be obtained if the circuit is built within a grounded, metallic enclosure, being careful to prevent coupling of the high-voltage AC output into the rest of the circuit.

LISTENING TO INDIVIDUAL PHOTONS

All of the experimental demonstrations above have looked at the particle nature of light using billions of photons. Wouldn't it be more convincing if we could observe a single particle of light at a time?

The setup of Figure 33 gives us an opportunity to detect individual photons. The detector is the PMT probe that we just built, and single photons are produced by dimming a laser through very dark photographic filters down to the level where only a few photons reach the photocathode every second.[67]

Let's work out the level of attenuation needed so that only one photon reaches the PMT in one second. We use a HeNe laser in our setup, which puts out photons at a wavelength $\lambda = 632$ nm. According to Einstein's proposition, each photon from our HeNe laser has an energy of:

$$E = hf = h\left(\frac{c}{\lambda}\right) = 6.626 \times 10^{-34} [\text{J} \cdot \text{s}] \times \left(\frac{2.998 \times 10^8 \left[\frac{\text{m}}{\text{s}}\right]}{632 \times 10^{-9} [\text{m}]}\right) = 3.143 \times 10^{-19} \text{J}$$

where c is the speed of light and λ is the wavelength of light (in this case 632 nm).

Our HeNe laser produces 5 mW of power, so we can calculate how many photons are emitted by the laser each second:

$$\frac{5 [\text{mW}]}{3.143 \times 10^{-19} \left[\frac{\text{J}}{\text{photon}}\right]} = \frac{5 \times 10^{-3} \left[\frac{\text{J}}{\text{s}}\right]}{3.143 \times 10^{-19} \left[\frac{\text{J}}{\text{photon}}\right]} = 1.591 \times 10^{16} \left[\frac{\text{photons}}{\text{s}}\right]$$

This means that we need to use very dark filters to drastically cut down on the number of photons that make it to the PMT. If we want only one photon per second to reach the PMT, we need to let only one photon out of 1.591×10^{16} photons through the filter. You could go to the beach on Venus with sunglasses that dark!

A dark photographic filter is called an *attenuator*, and the best kind for our purpose is a "smoked-glass" type known as a *neutral-density filter*. These attenuators weaken the beam with no significant dependence on wavelength. The darkness of a neutral-density filter is called *optical density* (D), which is defined as the base-10 logarithm of the desired attenuation.

Ideally, to have a single photon reach the PMT we need to attenuate the beam using a neutral-density filter with an optical density $D = \log_{10}(1.591 \times 10^{16}) = 16.20$. However, as Figure 33b shows, the RCA 6655A PMT will respond to only 1% of 632 nm photons impacting on its photocathode. In addition to the neutral-density filters, we added two narrow band-pass filters tuned to our laser's wavelength,

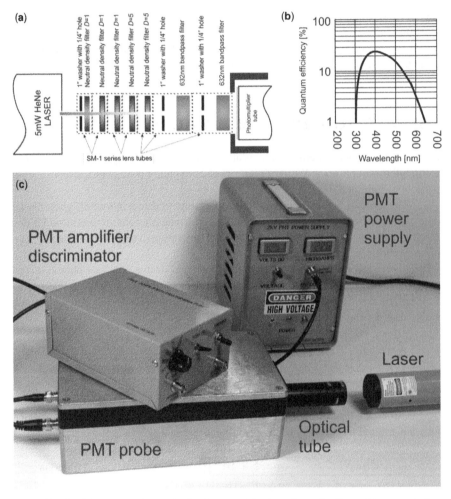

Figure 33   In this setup to detect single photons, (**a**) a laser beam is dimmed by very dark filters so that only a single photon reaches the PMT at any given time. (**b**) We used a helium–neon (HeNe) laser, but you can use any laser pointer by making the right adjustments to the system to match your laser's power and wavelength, including accounting for the PMT's quantum efficiency. (**c**) Our setup showing the high-voltage power supply (Figure 31) to bias the PMT, PMT probe (Figure 30) with the optical tube installed, the amplifier/discriminator (Figure 34), and the HeNe laser.

so that stray photons in the room cannot reach the PMT tube.[‖] Each of these filters allows only 45% of the desirable 632 nm photons through. We can thus recalculate the density needed to cause 1 photon/s to be absorbed by the photocathode in our setup:

$$D = \log_{10}\left(1\% \times 45\% \times 45\% \times 1.591 \times 10^{16}\right) = \log_{10}\left(3.22 \times 10^{13}\right) = 13.5$$

[‖]The laser-line band-pass filters may be omitted if you operate the setup in a dimly lit room.

There are other real-world effects that cause photon losses, so a good start is to use $D = 10$ and then work your way up to yield about one laser photon detection per second.

We purchased our neutral-density filters from Thorlabs and the laser-line filters from Edmund Optics (model 43133). We opted to standardize our optical component mounting on Thorlabs' series of SM lens tubes, so we bought our filters premounted inside SM1L03 cells. By the way, the total density of a stack of filters is simply the sum of their individual densities, so to get $D = 13$, we used two $D = 5$ and three $D = 1$ neutral-density filters.

Since the peak response of the RCA 6655A photomultiplier is close to 400 nm, you can get better quantum detection efficiencies if you have a green ($\lambda = 532$ nm) or violet ($\lambda = 405$ nm) laser pointer available. You will also need to change the filters to ones compatible with your laser's wavelength. Finally, you will need to recalculate the overall number of photons after the neutral-density filters, since the number of photons emitted at a set power depends on their wavelength. For example, a 5-mW green laser will produce "only" 1.34×10^{16} photons/s, or almost 16% fewer photons than a HeNe laser of the same power.

Figure 34 shows the circuit needed to turn the PMT detections into audible clicks. Despite the enormous gain of the PMT, the signals produced by our photomultiplier probe are very small when illuminated by just a few photons at a time. A *preamplifier* is thus needed to easily interface the probe with other instrumentation. Our PMT probe outputs a very fast,** negative-going pulse every time that a photoelectron is produced at the photocathode. Op-amp U1 is configured as an inverting *charge-sensitive amplifier*, which is one of the most common ways of detecting signals from a PMT's anode.

The positive-going pulses coming from the charge-sensitive preamplifier are turned into a Gaussian pulse by first differentiating (high-pass filtering) and then integrating (low-pass filtering) them using the circuits built around U2A and U2B. Trimmer R23 is used for pole cancellation to yield as clean a Gaussian pulse as possible through this simple configuration.

Pulses are amplified by U2C. The gain of this stage is selected via SW1. U3 is a comparator, which is used to convert pulses over a preset threshold into inverse-logic digital pulses that indicate a photon strike. This stage gets rid of small-amplitude pulses caused by thermal noise in the PMT, and is commonly known as a *discriminator* in physics instrumentation. Lastly, the pulses are stretched by one-shot U4 to provide an audible "click" through the piezo speaker. We added a buffer amplifier U2D so that the amplifier's analog signal may be observed on an oscilloscope. In addition, we provided an output from the discriminator to trigger other data acquisition equipment. We will use these outputs in a later experiment.

When in operation, this setup produces an irregular series of random clicks. Fine-tune the discriminator threshold so the clicking disappears when the laser is

---

**The pulses produced at the anode of the PMT by a single photon or short photon burst are negative-going and last for about 1 μs. For this reason, the op-amps selected for this application have very high bandwidth. If you need to substitute the MC34081 for a different model, select a JFET-input op-amp with at least 10-MHz bandwidth.

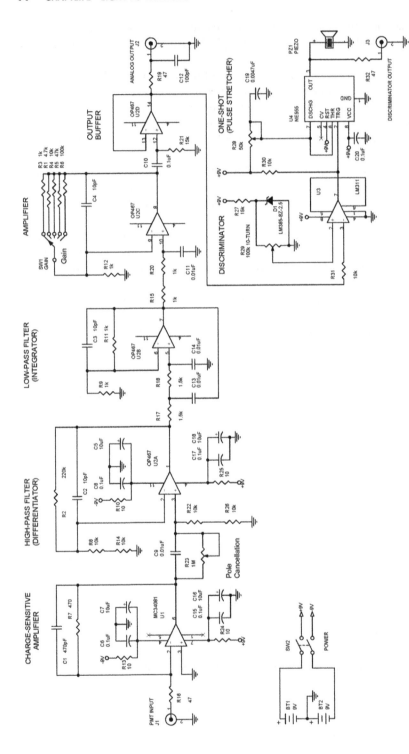

Figure 34 The PMT signal-processing circuit amplifies the narrow pulses detected by the PMT. The discriminator stage removes small pulses produced by thermal noise in the tube. A pulse stretcher outputs pulses that can be heard on a speaker.

off. Increasing the photon count to around 10 photons/s (by reducing the attenuation) makes for a signal that is much easier to detect. The output should sound like the clicking of a Geiger counter, and you may need to assure others within earshot that you are not playing with radioactive materials (yet).

Counting clicks while playing with attenuation should convince you that each count is indeed the result of a single photon being detected by the apparatus, demonstrating the existence of Einstein's quanta as individual particle entities.[††]

WHERE DOES THIS LEAVE US?

Einstein made very convincing arguments for the particle nature of light. However, as we saw in chapter 1, Young's experiments with the interference of light, and Maxwell's explanations about the polarization of electromagnetic waves had already destroyed Newton's view of light as a stream of particles. So where does this leave us?

We will return to this discussion in chapter 5. Suffice it to say that the question of whether light is a wave or a particle brought about the core philosophical shift that evolved into today's quantum physics: light behaves BOTH as a wave and as a stream of particles!

EXPERIMENTS AND QUESTIONS

1. Solder two temporary wires to the two electrodes of the lightbulb that you will use for the blackbody radiation experiments. Measure and record the room-temperature resistance (R_o) of the tungsten filament using a precision multimeter. Unsolder the wires and place the lightbulb on its base. Connect the lamp to the variable power supply and two multimeters so you can simultaneously measure the voltage between the lamp's terminals and the current through the filament. Increase the voltage in 1-V steps up to 14 V (or your lamp's rated voltage). Measure current at each voltage. Estimate and plot the filament's temperature at each voltage using:

$$T = 300°\text{ K} + \frac{\frac{\left(\frac{V}{I}\right)}{R_o} - 1}{4.5 \times 10^{-3}} [°\text{K}]$$

2. Use a setup constructed as shown in Figure 21 to measure the spectrum of the light emitted by the filament at three lamp voltages corresponding to temperatures of 3,800° K; 3,300° K; and 2,700° K. Correct your measurements according to the responsivity curve of your sensor (Figure 22b). Plot the intensity of the light as a function of wavelength. What happens to the spectrum of the blackbody radiation as temperature increases? Explain.

[††]"Photon bunching" does happen sometimes in an attenuated photon beam. For this reason, physicists who conduct rigorous single-photon experiments usually use a photon source that not only emits one photon at a time, but also provides confirmation of the photon's emission. We will learn about these entangled photon sources in Chapter 8.

3. Calculate the Wien peak of the blackbody spectrum at the three filament temperatures at which you obtained your data:

$$\lambda_{MAX} = \frac{b}{T} = \frac{2.898 \times 10^{-3}[m°K]}{T[°K]}$$

How well do the peaks in your data compare to Wien's estimate?

4. Do your experimental results match the spectral curve predicted by the expression developed by Rayleigh and Jeans? Explain.

5. How do astronomers measure star temperatures?

6. The peak of the solar spectrum is 502 nm. Estimate the temperature on the surface of the sun.

7. Use the setup of Figure 26 to measure the stopping voltage V_s for each wavelength. Plot the stopping voltages against LED frequency. Fit your measurements to a linear equation using linear regression. The slope of the graph you create is an estimate of h/e based on Einstein's equation for the photoelectric effect. Just as in the prior experiment, estimate Planck's constant h. How well does this estimate of h match the accepted value for Planck's constant?

8. Does the color of light affect the maximum energy of the electrons emitted from the photocathode? Explain.

9. Extend the line you obtained when graphing your data to the point where $V_s = 0$ to estimate the threshold frequency for photoemission for your phototube. Use an IR LED with a wavelength under the estimated threshold for photoemission to illuminate your phototube. What is your measurement of V_s for this LED? Explain your result.

10. Use a green LED in the setup of Figure 26. Power the LED at its nominal voltage and measure V_s. Reduce the LED supply *current* by 10% to reduce the intensity of the light. Measure V_s once again. Do the same after dropping the current a further 10%. What is the effect of varying the intensity of the light on V_s? Explain your results. Does the intensity of the light affect the maximum energy of the electrons emitted from the metal?

11. Use the same setup as before, but set $V_s = 0$. Measure the photocurrent at the three LED intensities (100%, 90%, and 80% of nominal LED current). Does the intensity of the light affect the rate at which the electrons are emitted from the metal? Explain.

12. Use the setup of Figure 27 to measure the time that it takes for the LED light to cause a photoelectric current. What is the worst-case time needed for the photoelectric current to appear after the LED is powered? How does this compare to the time expected for the photoelectric effect if the wave theory of light would apply? Provided you wait a sufficient amount of time, will an electron always be emitted when you shine light on a metal? Explain.

13. Connect the circuit of Figure 34 to the PMT probe. Power the PMT probe from a low-noise, low-ripple power supply (such as the one in Figure 31). Using the setup of Figure 33, adjust the discriminator threshold until you can hear the clicks produced by single photons releasing photoelectrons in the PMT. Count the number of clicks within 1 min and calculate the average time interval between clicks. If the speed of light is assumed to be 300,000,000 m/s, what is the average distance between two

successive photons causing clicks in the system? How does that distance compare to the dimension of the optical tube (the distance between the first neutral-density filter and the face of the PMT)? Using this average, and assuming that a single ND filter at the beginning of the tube would be used, what is the maximum number of photons present within the optical tube at any one time?

CHAPTER 3

ATOMS AND RADIOACTIVITY

Although physicists felt comfortable in the late 1800s with their capabilities to describe the macroscopic world, many questions remained unanswered. Lord Kelvin's lecture on the "Nineteenth-Century Clouds over the Dynamical Theory of Heat and Light," referred to two specific problems that interested him. Namely, he discussed the lack of physical theory to explain the speed of light, and the failure of physics to explain blackbody radiation. However, there were many other puzzles that were just as baffling.

One challenge was explaining the recently discovered phenomenon of radioactivity. Physicists were perplexed by the idea that energy could be spontaneously generated by matter. The classical view of physics simply couldn't explain how rocks could spontaneously glow in the dark or fog photographic film without light.

In 1896, French physicist Henri Becquerel learned about the greenish glow produced by recently-discovered X-rays in vacuum tubes. He decided to investigate whether there was any connection between the glow produced by X-rays and naturally occurring fluorescence found in certain minerals. Becquerel had inherited from his father some uranium salts, which fluoresce on exposure to UV light, and which were known to fog photographic plates, even if the plates were stored in dark envelopes. He hypothesized that uranium salts emitted X-rays when "charged" by exposure to sunlight. Much to his surprise, the salts fogged photographic paper, although the day he chose for his experiment was dark and rainy.

Later, Becquerel showed that the rays emitted by uranium were different from X-rays, because they could be deflected by electric or magnetic fields. For his discovery of spontaneous radioactivity, Becquerel was awarded half of the 1903 Nobel Prize for Physics. The other half was awarded to Pierre and Marie Curie for their study of "Becquerel radiation."

In this chapter, we'll take a step back in history and discuss atoms and radioactivity. Research in these fields became crucial for taking Planck and Einstein's quantum from being a mere curiosity in the study of light to its place in our understanding of matter. We will return to proper quantum physics in chapter 4.

THE NEED FOR VACUUM

The experiments in this chapter will deal with the production and detection of subatomic particles, such as electrons and nuclei. As these interact with matter, including

Exploring Quantum Physics Through Hands-On Projects. By David Prutchi and Shanni R. Prutchi
© 2012 John Wiley & Sons, Inc. Published 2012 by John Wiley & Sons, Inc.

air, many of the experiments need to be carried out inside glass tubes out of which the air has been pumped.

The normal atmospheric pressure at sea level is right around 1.0 atmosphere, or 760 Torr. The composition of air is approximately 78% nitrogen and 21% oxygen, with the remainder consisting of a mix of carbon dioxide, water vapor, and other trace gases. At atmospheric pressure, there are roughly 10^{19} of these molecules per cubic centimeter, which makes it very difficult for a subatomic particle to travel even a short distance without hitting an atom. The average distance that a particle can travel before colliding with other particles is known as the *mean free path*. However, this number depends very much on the particles involved. For example, at 1 Torr, an air molecule can go about 0.05 mm before colliding with another molecule, but an electron can travel about 1 cm before hitting an air molecule. At a pressure of 1 mTorr, these distances increase to 5 cm and 9 m, respectively.

The objective of running subatomic particle experiments in a vacuum is exactly to prevent collisions between the particles and air. The depth of the vacuum needed depends on many factors, including the type of particles being studied, the number of collisions that will not affect measurements too much, and the physical size of the apparatus.

Table 3 shows the conventional ranges into which vacuum levels have been divided. These ranges are not standard, and each technical organization or country

TABLE 3   Pressure within Various Vacuum Ranges[a]

Vacuum range	Examples	Pressure in Torr
Atmospheric pressure	Pressure at sea level	760
Low vacuum	Vacuum cleaner (600 Torr) Pressure at the summit of Mt. Everest (225 Torr) Pressure outside an airplane at cruising altitude (170 Torr)	760 to 25
Medium vacuum	Freeze-drying (1 Torr) **Cold-cathode glow-discharge tubes (0.01 to 0.1 Torr)** Incandescent lightbulb (0.01 Torr) Pressure at 90-km altitude, on the limit of space (10^{-3} Torr)	25 to 10^{-3}
High vacuum	Thermos bottle insulation (10^{-3} Torr) Vacuum insulation for the Large Hadron Collider (10^{-6} Torr) Commercial vacuum tubes ($<10^{-7}$ Torr) Pressure outside the International Space Station (10^{-8} Torr)	10^{-3} to 10^{-9}
Ultra high vacuum	Commercial CRTs (10^{-9} Torr) Pressure on the surface of the Moon (10^{-11} Torr)	10^{-9} to 10^{-12}
Extremely high vacuum	Large Hadron Collider Deep space	$<10^{-12}$

[a] The range we will use in our electron-beam experiments is shown in bold letters.

defines them differently. However, this table should give you an idea of the approximate levels used for different applications.

THE MECHANICAL VACUUM PUMP

We will conduct our experiments in the range of 10 Torr to 10 mTorr, which is easy to achieve using a low-cost, mechanical vacuum pump. Very good results have been obtained by amateur experimenters and educators using vacuum pumps made for servicing air conditioning and refrigeration units. These vacuum pumps are not part of the refrigeration system itself, but rather are used by service technicians to reduce the internal system pressure of a refrigeration/air conditioning system so moisture and other contaminants can be removed.

A robust, relatively low-cost unit that is popular with experimenters is the model 15600 refrigeration service pump made by Robinair. It is rated to reach a vacuum of 20 mTorr (refrigeration vacuum pumps are commonly specified in terms of their base pressure in "microns," where 1 micron = 0.001 Torr = 1 mTorr). In practice, we reach pressures down to 1 mTorr by using Kurt J. Lesker TKO 19 Ultra vacuum oil instead of the Robinair oil recommended for this pump.

The vacuum intake port of the Robinair pump is terminated with a 45° flared fitting (a standard of the Society of Automotive Engineers, or SAE). SAE flared fittings seal by the mating of two beveled metal surfaces. The function of the threads is simply to draw these two surfaces together. These couplings are usually used for low-pressure applications, such as refrigerant and fuel lines in conjunction with copper tubing, which flares easily to 45°. This is not the most convenient connection for our purposes, so visit the hardware store and put together an adapter that will take you from the $\frac{1}{2}$-in. SAE flare to a hose barb that you can connect to $\frac{1}{2}$-in. ID steel spring–reinforced PVC hose, or better yet, to a vacuum-service rubber hose from the Kurt J. Lesker company. Remember to buy any copper flare gaskets and O-rings that you may need to make vacuum-tight connections.

The other side of the hose will connect to your vacuum chamber. The best way is through the type of *vacuum flange* that conforms to one of the industry standards to connect vacuum chambers, tubing, and vacuum pumps to each other. We prefer to use a standard quick-release flange known as a quick flange (QF) or Klein flange (KF). KF/QF flanges* are made with a chamfered back surface that attaches with a circular clamp and an elastomeric O-ring mounted in a metal centering ring. These come in various sizes that are indicated by their inner diameter in millimeters. We use mostly KF16 (16-mm ID) flanges, some KF25 flanges, and occasionally KF40 flanges.

As such, you will need to terminate the hose with a KF-to-hose adapter to fit your hose. For example, you could use a KF16 to $\frac{1}{2}$-in. hose adapter made by the Kurt J. Lesker Company (catalog number QF16-050-H). Your vacuum components would then attach to the KF16 flange using another KF16 flange, a centering ring

---

*KF and QF are only designations. A component labeled "QF" is directly compatible with one of the same size labeled "KF."

(e.g., Kurt J. Lesker catalog number QF16-075-ARB) and a clamp (e.g., Kurt J. Lesker catalog number QF16-075-C). Use some vacuum grease (e.g., Fomblin VAC3 grease, or Dow Corning high-vacuum grease) on all joints and rubber O-rings.

As an alternative, Steve Hansen, editor of the amateur vacuum experimenter's magazine *The Bell Jar*, sells a ready-made inlet manifold for the Robinair pump. A vacuum gauge and vacuum chamber can be attached directly to this manifold. The bottom connector is a $\frac{1}{2}$-in. SAE flare adapter that mates directly to the pump's vacuum port. The manifold has one side port fitted with a needle valve that can be

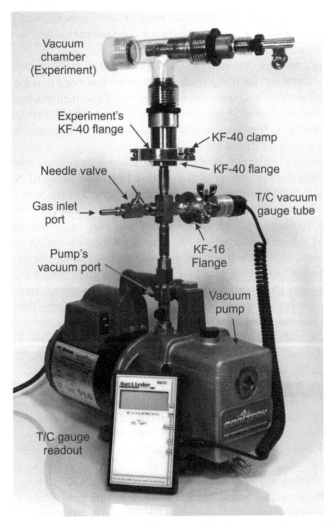

Figure 35   Our simple vacuum setup, which reaches 10 mTorr, comprises a refrigeration service vacuum pump, a vacuum manifold handcrafted by Steve Hansen, and a T/C vacuum gauge. Our experimental chambers connect to the manifold through a KF-40 flange. A pinch valve connected to the manifold allows us to introduce small amounts of gas into the system.

used for pressure control and inert gas inlet. A second side port connects to a standard thermocouple (T/C) gauge tube (more on those in the next section). As shown in Figure 35, the main port of the manifold is a KF40 flange that attaches to the vacuum chamber used for an experiment. The pump does vibrate quite a bit, so you will need to rest the pump on some vibration-absorbing material if your vacuum chamber is connected to the pump through a rigid coupling. We use a mat made of Sorbothane®, but any other shock-absorbing material will do. For example, you could use the gel pads from insoles sold at pharmacies to reduce impact in walking shoes.

THE VACUUM GAUGE

Simple mechanical pressure gauges don't usually work at pressures below 10 Torr, so we need a different way of measuring pressure in our vacuum systems. We use the most inexpensive pressure sensor that works at these levels, which is the *T/C gauge*. It operates by measuring the thermal conductivity of the gas inside the chamber. As shown in Figure 36, the T/C tube contains a filament heated with an AC constant current and a T/C in contact with the filament. As the pressure decreases, the filament becomes hotter, because the number of gas molecules hitting the wire and conducting heat away from the wire decreases. As temperature rises, the T/C voltage increases and is measured by a sensitive meter that has been adjusted according to the T/C tube's calibration curve.

Thermocouple gauge readout units (sometimes called T/C gauge controllers) are easy to come by on the surplus market. Make sure that the one you buy is compatible with the T/C gauge tube you intend to use. Alternatively, you may build your own gauge readout, as shown in the schematic diagram of Figure 37. This circuit is designed for the Teledyne Hastings type DV-6M T/C gauge. We use the model that is terminated with a KF-16 flange. The resistance of the heater filament is 18 Ω (don't test this with an ohmmeter because you will burn the filament). It requires 0.38 VAC to operate at a current of 21 mA. The T/C's output is 10 mV at high

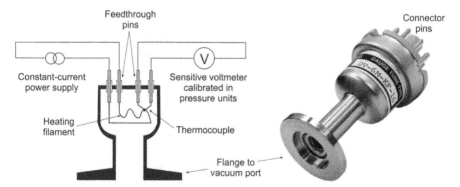

Figure 36 A T/C vacuum gauge is able to measure pressures in the range of 1 to 10^{-3} Torr. It operates by measuring the thermal conductivity of the gas inside the chamber. As the pressure drops, the filament becomes hotter, causing the T/C voltage to increase.

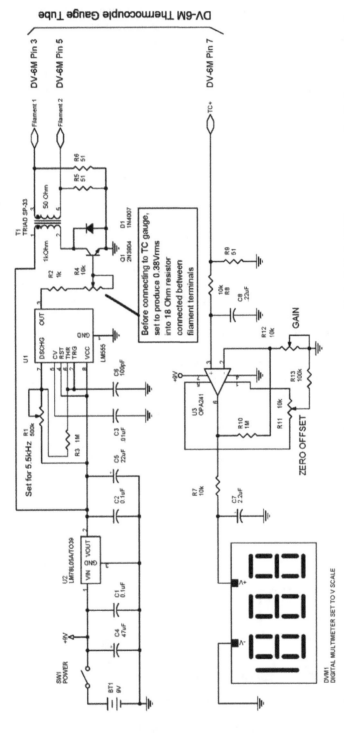

Figure 37   This is a simple readout circuit that you can build to measure pressure with the DV-6M T/C vacuum gauge tube. Before connecting the T/C tube, make sure that the transformer's output across an 18 Ω resistor is 0.38 V$_{RMS}$.

vacuum, and drops as pressure increases (see Figure 38). The DV-6M is useful in the range of 1 mTorr to 1 Torr, but is most sensitive between 10 and 200 mTorr. The filament is connected between pins 3 and 5, but the T/C and heater functions are combined in this tube (unlike the "textbook" diagram of Figure 36), so only one connection (pin 7) carries the T/C's output signal.

In the readout circuit of Figure 37, a 555 timer IC controls transistor Q1, which drives the primary of a small audio transformer T1. The output of this transformer is used to heat the DV-6M gauge tube's filament. An op-amp (U3) is used to amplify the signal from the T/C so that it can be measured with a voltmeter. Before connecting the gauge tube, connect an 18 Ω resistor at the transformer's output and adjust the voltage across this resistor to exactly 0.38 V_{RMS}.

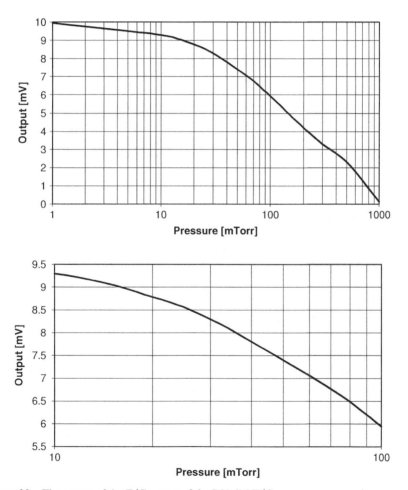

Figure 38   The output of the T/C output of the DV-6M T/C vacuum gauge varies as a function of pressure. These values are valid when the gas in the vacuum chamber is air, and the current into the filament is calibrated to produce 10 mV at 0.01 mTorr. You can use these scales to calibrate your homemade T/C gauge readout unit.

A VERY-HIGH-VOLTAGE POWER SUPPLY

The experiments in this chapter will also require us to use very high voltages—well in excess of the 2,000 V that can be obtained from the power supplies we built to power the PMT probe. Fortunately, precise regulation at these voltages is not needed, so a high-voltage power supply that can produce over 100,000 V is easy to build. Figure 39a shows a DC-to-AC inverter that is used to drive the high-voltage multiplier of Figure 39b. In this power supply, a push–pull oscillator drives a TV flyback transformer from an old color TV.[†] The original primary of the flyback is not used. Instead, new primaries are made by winding two sets of four turns each of insulated #18 wire around the exposed core of the flyback transformer. Feedback for the oscillator is obtained through an additional coil of 4 turns of #24 wire wound around the core. The application of 12 V at the input of the flyback driver should produce 100 to 200 kVDC (depends on the flyback used) at the output of the flyback's quintupler.

The output polarity of the so-called *Cockroft–Walton multiplier* depends on the way in which its diodes are oriented. Since some experiments call for both polarities, we designed the multiplier to yield either positive or negative output. If the high-voltage AC output of the flyback is connected to point A of the voltage multiplier, and point B is connected to ground, then the output at point D will be positive. If however point C receives the high-voltage AC, and point D is connected to ground, then point B will be negative. The multiplier should be built on a piece of clean perforated board suspended by nylon spacers inside a plastic container. Banana connectors may be installed on the plastic container and connected directly to points A, B, C, and D. The plastic container should then be filled with pure mineral oil (can be purchased at a pharmacy) to completely submerge the multiplier circuit assembly, which prevents high-voltage breakdown between components. Please note that this is a dangerous device! It produces high voltages that can cause very painful or lethal electrical shocks. In addition, spark discharges can be produced that can ignite flammable materials or volatile atmospheres.

A VACUUM TUBE LEGO® SET

Making vacuum tubes from scratch is almost a lost art. It involves plenty of practice in technical glassblowing, as well as an understanding of materials and vacuum techniques. However, vacuum tubes that operate at relatively high pressures and require continuous pumping are easy to build by cobbling together glass tubes, copper piping, and rubber corks. We prefer to use a more upscale version of this technique and follow the lead of vacuum expert Steve Hansen, who uses a neat LEGO set of interlocking glass pipettes and bulbs made by Ace Glass to construct experimental

---

[†]"Old" is the important word. Most new TV flybacks have an integrated rectifier/tripler. Our power supply requires access to the raw secondary high-voltage terminal, as was common in tube color TVs. Flyback transformers of this kind are available from Information Unlimited. In fact, you could look into buying one of their GRADRIV1 kits or ready-made high-frequency, high-voltage power supplies if you would like to skip building the circuit of Figure 39a.

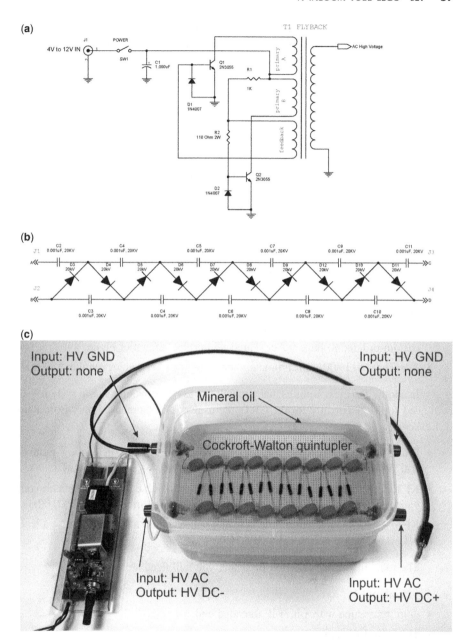

Figure 39   Schematic diagram of high-voltage power supply. (a) A push–pull oscillator drives a TV flyback to produce AC high voltage. (b) A quintupler is built separately and operated while dipped in pure mineral oil. Terminal B is the negative output if C is connected to the flyback's AC high-voltage terminal and D to ground. (c) The quintupler is shown (with cover removed) connected to an Information Unlimited GRADRIV1 to produce over 50 kVDC from a 15-V input.

vacuum tubes.[8,9] The glass components and connectors made by Ace are not designed for vacuum applications. Ace makes them for chemistry, air sampling, and chromatography analysis. However, they work really well in the relatively modest vacuum ranges needed to reproduce these early electron-tube experiments.

The set of Ace Glass components that you'll need to conduct the experiments in this book includes:

- Ace-Thred #25 "T" connecting adapter (Ace catalog number 5829-12)
- 2 Ace-Thred #25 connectors (Ace catalog number 7644-20)
- Ace-Thred #11 connector (Ace catalog number 7644-10)
- 2 Ace-Thred #25 couplings (Ace catalog number 5841-16)
- 2 Ace-Thred #25 bushings (Ace catalog number 7506-10)
- 7 Ace-Thred #7 bushings (Ace catalog number 5029-10)
- 3 Ace-Thred #7 plugs (Ace catalog number 5846-44)
- 2 Ace-Thred #11 bushings (Ace catalog number 7506-02)
- Ace 1,000 mL Gledhill flask with Ace-Thred #25 connector (Ace catalog number 14205-09)
- Ace-Thred #25 12-in. column (Ace catalog #7488-24 air-sampling manifold) custom-ordered to have 6 Ace-Thred #7 ports as shown in Figure 47b
- Ace-Thred #11 300-mm chromatographic column (Ace catalog number 5820-04)

In addition to the couplings that you will use as part of your vacuum system, you will need some specifically for the CRTs themselves. We purchased ours from Kurt J. Lesker.

- Brass QuickConnec coupling for $\frac{3}{4}$-in. tube (Kurt J. Lesker catalog number B-075-KM)
- KF to QuickConnec coupling for 1-in. tube and a KF connector to match your vacuum manifold (we use a Kurt J. Lesker catalog number QF40XVC100)
- KF to $\frac{3}{8}$-in.-OD tube half nipple to match your vacuum manifold (we use a Kurt J. Lesker catalog number HN-SPL219)

The rods, tubes, and assorted parts that you will need are:

- $\frac{3}{8}$-in. aluminum rod
- $\frac{7}{16}$-in. brass tube with $\frac{1}{4}$-in. FIP threaded end
- Two $\frac{1}{4}$-in. FIP caps
- Two 1-in.-OD × 2-in.-long stainless steel tubes (made from Kurt J. Lesker catalog number SST-0100I tube, which is sold by the inch)
- Two $\frac{3}{8}$-in. ID shaft collars with setscrew (McMaster-Carr catalog number 6166K23)
- Seven $\frac{1}{4}$-in. ID shaft collars with setscrew (McMaster-Carr catalog number 9414T6)

- $\frac{1}{4}$-in. × 2-in.-long stainless steel, internally-threaded spacers for 4-40 screws (McMaster-Carr catalog number 91125A473)
- Brass screws (McMaster-Carr catalog number 92480A106)
- Brass sheet (0.015-in. thick, K&S Engineering)
- 1-in. ID vacuum rubber hose (Kurt J. Lesker catalog number T100)

PHOSPHOR SCREENS

Electron beams in a good vacuum are invisible, but a screen coated with a material that fluoresces when hit by electrons can be used to make the beam visible. These fluorescent materials are called *phosphors*, and are used most commonly in cathode-ray oscilloscope and TV screens to produce an image by steering an electron beam inside the CRT. Phosphors don't contain the element phosphorous. They are usually fine powders of zinc sulfide (ZnS) that have been "activated" with a tiny bit (a few parts per million) of either copper or silver. The ZnS(Ag) or ZnS(Cu) powders are then mixed with a binder (an adhesive to help them stick to a surface) into a suspension and deposited onto the screen. Another popular phosphor for electron-tube screens is zinc silicate activated with manganese.

We made a screen for our Ace Glass set of components from an Ace 1,000 mL Gledhill flask (catalog number 14205-09), the base of which we coated with a phosphor (Figure 40). The bottom of the flask is not perfectly even, so we first poured 20 mL (around $\frac{3}{4}$ oz.) of clear, slow-cure epoxy on the flask and let it sit for 3 days to form a flat, transparent surface.[‡]

After that, we thoroughly mixed 3 g of phosphor powder sold for recoating electron-microscope screens (made by Structure Probe, as their catalog number 04129-AB) with 0.75 g of Beacon 527 multi-use glue (from a craft store) and 10 g of acetone (from the paint thinner aisle in a hardware store). We then poured the suspension onto the bottom of the flask (on top of the cured epoxy), swirled it for even distribution, and allowed it to sit undisturbed until the acetone completely evaporated. We also painted a 1-cm wide fluorescent stripe on the side wall of the flask between the screen and the Ace-Thred #7 side port of the Gledhill flask.

Commercial phosphors made for electron microscopes are somewhat expensive, so you could try other materials that fluoresce under electron bombardment. These include many highlighter pen inks, UV pen and stamp-pad inks, and glow-in-the-dark paints.

An alternative to making your own screen is to use one from an oscilloscope CRT that has been cut away and matched to an Ace-Thred connector. As shown in Figure 40, we cut a 3HP7 oscilloscope CRT in half and mated the screen to an Ace Glass #25 threaded connector (Ace catalog number 7644-20), using a piece of 1-in. ID rubber hose (Kurt J. Lesker catalog number T100 coated on the inside with Dow Corning high-vacuum grease). We tried to cut the neck of the CRT with many of

---

[‡]Just after pouring the epoxy we evacuated the flask down to 100 Torr for 10 minutes to pull out bubbles from the epoxy.

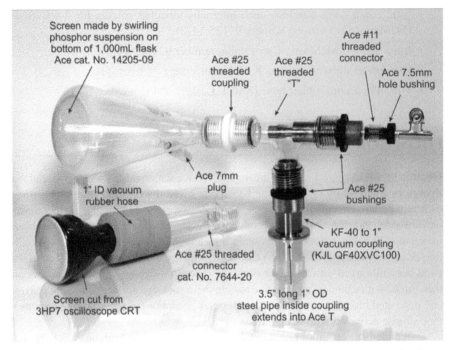

Figure 40   We use two different screens for our homemade CRTs. We built one out of the screen of an oscilloscope CRT that we cut in half. The other is an Ace Glass flask into which we poured a suspension of ZnS phosphor. Both tubes can be connected to our apparatus through #25 Ace-Thred connectors.

the common bottle-cutting methods (hot wire, score with diamond cutting wheel, etc.), but had the best results cutting it with a Dremel rotary tool using a diamond grind disk from Micro-Mark (catalog number 82259 or 84611).

THE ELECTRON GUN

Our CRTs require a beam of electrons, which we produce using the assembly shown in Figure 41. Electrons are stripped off the cathode by a process known as *glow discharge* between an aluminum cathode rod and a hollow anode. The electron gun is built inside an Ace-Thred #11 connector (Ace catalog number 7644-10). The cathode is a $\frac{3}{8}$-in. aluminum rod sealed against the Ace-Thred #11 connector through an Ace-Thred #11 bushing. The anode electrode is a $\frac{7}{16}$-in. brass tube threaded to accept a $\frac{1}{4}$-in. FIP brass cap. An easy way of making it is from a $\frac{1}{4}$-in. brass nipple turned to reduce its body diameter to $\frac{7}{16}$-in. Electrons exit the gun in a fine beam from a $\frac{1}{32}$-in. hole drilled in the $\frac{1}{4}$-in. FIP anode brass cap. The gun couples to our Ace-Thred #25 tubes through an Ace-Thred #25 bushing. The gun is sealed against the bushing through a Kurt J. Lesker B-075-KM vacuum coupling that we epoxied to a 1-in.-OD $\times$ 2-in.-long stainless steel tube.

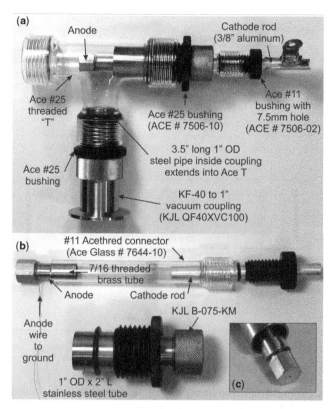

Figure 41　The electron gun for our homemade CRT (Figure 40) uses glow discharge between a cathode rod and a hollow anode. (**a**) The electron gun is built inside an Ace-Thred #11 connector (**b**) coupled to an Ace-Thred #25 bushing through a Kurt J. Lesker B-075-KM vacuum coupling epoxied to a 1-in.-OD × 2-in.-long stainless steel tube. (**c**) We drilled a $\frac{1}{32}$-in. hole in the $\frac{1}{4}$-in. FIP anode brass cap to allow a thin electron beam to exit the gun.

THE DISCOVERY OF THE ELECTRON

Some of the most important experiments to explain radiation were conducted in 1897 by British physicist J. J. Thomson. These experiments led Thomson to the discovery of the electron, for which he received the 1906 Nobel Prize in Physics. The story starts in 1857, when German physicist and glassblower Heinrich Geissler pumped the air out of a glass tube fitted with wires at both ends. The glass tube would produce glowing streamers when high voltage was applied across the electrodes—something that Victorians found extremely interesting and entertaining. Let's build our own tubes to see these effects firsthand, and ultimately understand the reasoning that Thomson employed in his discovery of the electron!

　　Our basic glow-discharge tube is shown in Figure 42a. It consists of a 30-cm-long Ace-Thred #11 glass tube (Ace catalog number 5820-04) that is terminated by a $\frac{3}{8}$-in. aluminum rod at one end and by a $\frac{3}{8}$-in. tube on the other end. Each of these

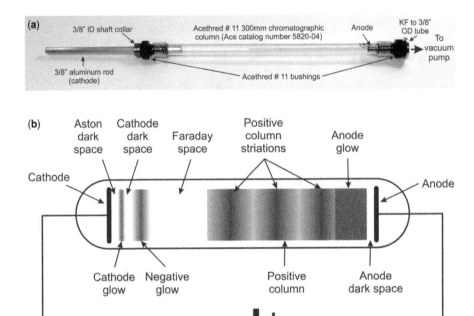

Figure 42   A glow-discharge tube simply requires two electrodes sealed within an evacuated glass tube. (**a**) Our tube consists of a 30-cm-long Ace-Thred #11 glass tube terminated by a $\frac{3}{8}$-in. aluminum rod at one end and by a $\frac{3}{8}$-in. tube on the other end, which connects directly to our vacuum manifold. (**b**) As the pressure in the tube is lowered, many interesting features appear in the glow discharge between the electrodes.

is sealed by an Ace-Thred #11 bushing (Ace catalog number 7506-02). The tube connects directly to the vacuum manifold (we use a KF-16 to $\frac{3}{8}$-in.-OD tube half-nipple, Kurt J. Lesker catalog number HN-SPL219). The aluminum rod is prevented from being sucked into the tube by a $\frac{3}{8}$-in. ID shaft collar. The negative output of the high-voltage power supply of Figure 39 connects to the aluminum rod, while the ground terminal connects to the vacuum port.

Nothing happens at atmospheric pressure, since the voltage of the power supply is too low to bridge through approximately 25 cm of air inside the column. However, a glow discharge in the form of a thin streamer forms between the electrodes once the pump is turned on and the pressure drops to around 100 Torr (roughly one-tenth of an atmosphere). As the pressure continues to drop, the streamer becomes more vivid, radiating with a pretty pink/violet glow.

When the pressure is lowered to around 5 Torr, the pink glow pulls away from the cathode, leaving a bright bluish glow next to the cathode and a dark space (known as the Faraday space) between the pink and blue light areas. At around 1 Torr, the blue negative glow separates from the cathode. At the same time, the pink positive column breaks up into a series of separate stripes (striations). These glow phenomena captivated the attention of physicists and noblemen of the time. Geissler had developed a mercury vacuum pump capable of evacuating tubes down to around 2 Torr, making it possible for him to produce and sell his Geissler tubes to universities, as well as for home entertainment.

By 1870, British physicist Sir William Crookes was using an improved vacuum pump to draw an even better vacuum out of a Geissler tube, and noticed a fluorescent glow in the glass close to the cathode electrode. The fluorescence of the glass made him conclude that some kind of ray was being emitted by the cathode, causing the minerals in the glass to glow. Crookes also observed that the glow inside the tube didn't start right at the cathode, but that there was a "dark space" between the metallic electrode and the glowing column. Furthermore, as he pumped more air out of the tube, the dark space grew toward the anode, until the tube was totally dark. At that point, fluorescence of the glass appeared all the way close to the anode.

As you will notice when you conduct your own experiments, the glass walls of the tube start to fluoresce at around 500 mTorr under the bombardment of rays within the tube. You can steer these rays and the fluorescence they cause to different parts of the tube's wall by bringing a strong magnet close to the tube (just be careful of getting too close to the high-voltage cathode!).[§]

The nature of these rays was a mystery in the 1870s. One possibility was that they were a type of wave, perhaps one of Maxwell's electromagnetic waves, just like light, but at a different wavelength. Another possibility was that these *cathode rays* were made of some kind of material particle. This latter view was supported by the fact that waving a magnet near the glass would move the rays around, causing the presumably negatively charged particles to be pushed around.

CATHODE-RAY TUBES

In 1879, Crookes had reached the conclusion that cathode rays must be particles of some sort, observing that they traveled in straight lines and were stopped by metallic objects in their path. As a demonstration, Crookes inserted an electrode shaped like a Maltese cross in the tube, and it cast a sharp shadow on the fluorescence at the end of the tube (Figure 43a). Maltese cross tubes are still commonly demonstrated in university physics classes. As shown in Figure 43b, we built our own version using a more jovial target.

We soldered a thin piece of brass to a $\frac{3}{4}$-in. copper washer. We cut out the happy face on the brass sheet, and soldered a brass screw to a tab we left on the brass sheet. We painted the happy face with some leftover phosphor so we can see when the

[§]Our vacuum pump is able to pump down much more deeply than the 250 mTorr or so that scientists could achieve in the 1870s. As you keep on pumping, you will see the Crookes dark space appear at around 250 mTorr, and it will expand greatly into the positive column at 100 mTorr. Eventually, the current through your tube will stop as the pressure drops even further, and the glow will vanish. Just before that happens however—at around 20 mTorr—X-ray emission begins, causing the glass walls to fluoresce again. If you have a Geiger counter available (see Figure 57), place the Geiger–Müller tube about 50 cm away from the glow-discharge tube and pay attention for a sudden increase in the number of counts.

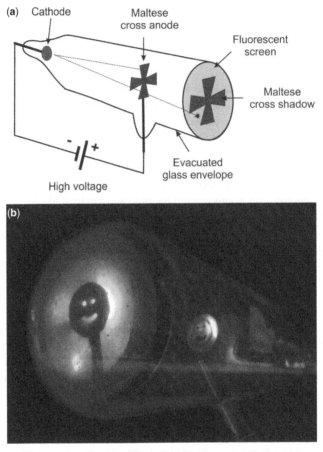

Figure 43   In 1879, Sir William Crookes inserted a Maltese cross–shaped anode into a CRT (a). The shadow on the fluorescent screen demonstrated that cathode rays travel in straight lines. Crookes also saw this as evidence that cathode rays must be some sort of particle that does not penetrate metal. (b) Our version of the Maltese cross tube uses a more jovial subject to cast a shadow.

cathode-ray beam hits it. We then placed the happy face inside the Ace 1,000 mL Gledhill flask (Ace catalog number 14205-09) on which we had deposited a fluorescent screen (Figure 40). We carefully inserted the brass screw into a threaded $\frac{1}{4}$-in.-OD × 4-in.-long spacer (made by joining two $\frac{1}{4}$-in.-OD × 2-in.-long spacers) that we introduced through the Ace-Thred #7 port on the Gledhill flask.

We coupled the flask to the electron gun of Figure 41 with an Ace-Thred #25 coupling (Ace catalog number 5841-16). We connected the negative high voltage of our power supply (Figure 39) to the gun's cathode (the $\frac{3}{8}$-in. aluminum rod), and the ground terminal to the vacuum manifold and the happy face support.

As you can see in Figure 43b, the shadow produced on the fluorescent screen is very sharp, indicating that the cathode rays travel in a straight line from their source (the small hole in the anode cap) to the screen. They are intercepted very cleanly by the happy face mask. Moving a magnet about the CRT deflects the cathode rays. Depending where the magnet is placed, it may shift the illumination of the mask, or distort the shadow on the screen.

Crooke's argument did not convince everyone. Most notably, it was rejected based on the results of an experiment conducted by Hertz in 1891. Hertz reasoned that the cathode rays should be deflected by an electric field if they were beams of negatively charged particles. Hertz placed two electrode plates within a CRT, and expected the beam to deviate when he applied a high voltage between the plates. The beam didn't move. The cathode rays were not deflected in the way that would be expected of electrically charged particles. We know today that cathode rays are deflected by an electric field, and that the reason Hertz didn't detect the deflection was the vacuum in his tubes was not sufficiently low to prevent inter-actions with the gas. However, Hertz didn't know this, and accepted his experimental results.

In a different experiment, Hertz placed a thin metal foil in the path of the rays, yet still saw some glow in the fluorescent screen behind it. Unknowingly, Hertz was pro-ducing X-rays that excited the screen's phosphor. The incorrect interpretation of these experiments convinced Hertz that cathode rays were not streams of particles, but rather some type of wave.

THOMSON'S FIRST 1897 EXPERIMENT—NEGATIVE CHARGE AND RAYS ARE JOINED TOGETHER

Hertz' (mistaken) finding that cathode rays were not deflected by an electric field was a true mystery, given that cathode rays were so easily bent by magnetic fields. In 1895, French physicist, and later Nobel Prize laureate, Jean Perrin built a CRT to investigate whether or not cathode rays transported charge.[10] A diagram of Perrin's original tube is shown in Figure 44a. His idea was to have cathode rays flow between the cathode and anode electrodes. However, a hole in the anode would allow some of the cathode rays to go into the hollow anode tube, where they would be probed by a collecting elec-trode. The very fine gold leaves on an electrometer would spread apart if the cathode rays conveyed charge to the collection electrode. This is because the leaves repel each other when they acquire similar electric charges. Their separation is a direct indication of the net charge stored on them. In addition, Perrin hypothesized that the collected charge should disappear as soon as the source of the radiation was removed if the cath-ode rays were some form of electromagnetic radiation. The fact that charge remained on the electroscope supported his hypothesis that cathode rays were made of electri-cally charged matter.

Perrin's experiment had shown unequivocally that cathode rays transported negative charge. However, not everyone was convinced that this was not an incidental effect of the radiation that produced fluorescence in screens. In his first 1897 critical experiment toward the discovery of the electron, British physicist J. J. Thomson built a modified version of Perrin's 1895 tube, as shown in Figure 45a. Thomson

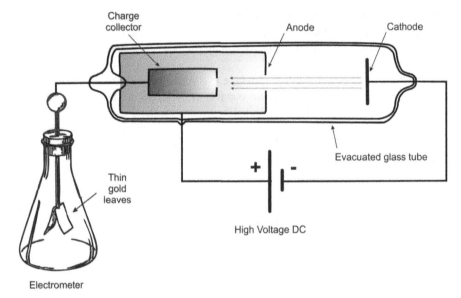

Figure 44   In 1895, French physicist Jean Perrin set out to investigate whether or not cathode rays transported charge. His CRT allowed some of the cathode rays to go into the hollow anode tube, where they would be probed by a collecting electrode.¶ The very thin leaves on an electrometer would spread apart if the cathode rays conveyed charge to the collection electrode.

wanted to see if, by bending the rays with a magnet, he could separate the charge from the rays. Without a magnetic field present, the cathode rays followed a straight line and produced fluorescence in the glass just in front of the anode. As expected, no charge was detected by the electrometer connected to a collection electrode tilted away from the ray's path. When a magnet was placed close to the tube, the cathode rays were bent toward the collection electrode, causing the electrometer to indicate charge. Critically however, the site of fluorescence also deviated as the magnet was brought close to the apparatus, until the fluorescence disappeared at the collector electrode. As Thomson saw it, the negative charge and the cathode rays must somehow be connected.

For our version of Thomson's tube, we simply retracted the "happy face" electrode on the side port of the flask of Figure 43 so that it would lie flat against the flask wall, as shown in Figure 45b. In addition, we connected a digital multimeter with 10 MΩ input impedance between the happy face electrode's post and ground. We set the multimeter to measure volts DC, essentially turning it into a sensitive microammeter in which the current in microamperes is equal to the voltmeter's reading divided by 10.

When you conduct the experiment, first evacuate the tube to approximately 100 mTorr and then turn on the high voltage. As you continue pumping, the cathode rays should produce a clearly visible mark on the screen at a spot across from the exit hole

¶Figure adapted from Perrin's original paper "New Experiments on the Kathode Rays," which was read before the Paris Academy of Sciences, December 30, 1895. A translation was published in *Nature*.[10]

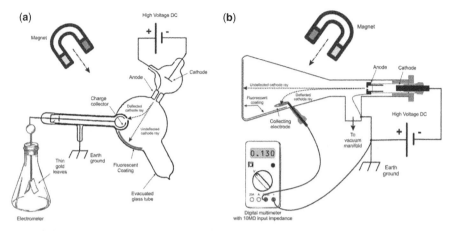

Figure 45 In 1897, British physicist J. J. Thomson conducted the first of three critical experiments that led him to the discovery of the electron. **(a)** He used a modified version of Perrin's tube to determine if the charge and radiation elements of cathode rays were different aspects of the phenomenon. As expected, the undeflected ray produced fluorescence and no charge detected by the electrometer. When a magnet was placed close to the tube, rays were bent toward the collection electrode, causing the electrometer to indicate charge. Critically however, the site of fluorescence also deviated as the magnet was brought close to the apparatus until the fluorescence disappeared at the collector electrode. We adapted this figure from Thomson's original 1897 paper on cathode rays.[11] **(b)** Our version of the experiment uses a digital voltmeter to detect charge deposited onto the collecting electrode.

of the electron gun's beam. Stabilize the pressure and voltage to maintain that spot on the screen. Measure the voltage on the digital multimeter. Now bring a magnet close to the tube, such that the cathode rays are bent toward the happy face electrode. This task will be easy if you followed our suggestion of painting a 1-cm fluorescent stripe on the side wall of the flask between the screen and the side port. The reading on the multimeter should increase quite dramatically when the cathode rays are deflected to fall on the collection electrode. Your voltmeter will let you determine the charge conveyed by the cathode rays (try inverting the connections to the multimeter), as well as the amount of charge arriving at the collection electrode (remember that $I[A] = Q[\text{Coulomb}]/\text{s}$).

THOMSON'S SECOND EXPERIMENT—ELECTROSTATIC DEFLECTION OF CATHODE RAYS

For his second crucial experiment, Thomson reattempted to deflect the cathode rays, much like Hertz had unsuccessfully tried before, but this time pulling a deeper vacuum in the tube. As shown in Figure 46, the cathode rays in his tube were made to pass between two parallel aluminum plates. Contrary to Hertz's experience, the cathode rays indeed deflected when a voltage was placed across the plates. Thomson showed that Hertz was unable to detect the deflection in his tube because the

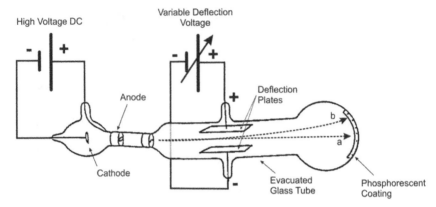

Figure 46   For his second crucial experiment of 1897, Thomson built a CRT that he evacuated very deeply. The cathode rays in his tube traveled between two parallel aluminum plates. As expected, the beam would hit the fluorescent screen with no deviation (**a**) when the plates were not electrostatically charged. However, applying a voltage between the plates deflected the beam (**b**) in an amount proportional to the potential difference between the plates. We adapted this figure from Thomson's original 1897 paper on cathode rays.[11]

vacuum was not sufficiently deep.[‖] However, the deflection could be easily detected in Thomson's tube, even in a weak electric field, provided the vacuum was good enough. Moreover, Thomson found that the deflection was proportional to the potential difference between the plates.

The tube pressure needed to observe electrostatic deflection is well within the capabilities of our humble refrigeration-service pump. In fact, we even need to leak a bit of air on purpose in order to keep the pressure at around 33 mTorr to keep the glow discharge going.

As shown in Figure 47, we added electrostatic deflection plates to our homemade CRT by introducing an Ace Glass segment with two sets of #7 ports placed at 90° to each other. We cut deflection plates from a thin brass sheet (purchased at a hobby store) on which we soldered a small brass screw. We keep the plates in place and bring out electrical connections through internally threaded $\frac{1}{4}$-in.-diameter × 2-in.-long stainless steel spacers. A rubber O-ring and a #7 Ace-Thred bushing seal each cylindrical spacer against its port. To deflect the beam, we connected one of the plates in the vertical pair to ground (the vacuum pump manifold). We also connected the ground terminal of our PMT bias power supply to the same point. We connected the other plate to the high-voltage terminal of the PMT power supply, and caused a deflection of the beam in the direction of the positively charged plate by increasing the voltage of the PMT power supply. When you conduct this experiment, try correlating the amount of deflection that you see on the screen against the voltage

‖Hertz was not able to detect deflection in his poorly evacuated CRT due to an effect known as *gas focusing* or *ionic focusing*. When the electron beam moves through the gas present in the tube, gas ions are produced that form a positive channel down the center of the tube, confining the electron beam to that channel. The electrostatic field used by Hertz was not sufficient to overcome gas focusing within the tube, and he therefore was not able to detect the deflection enacted on the electrons by an electric field.

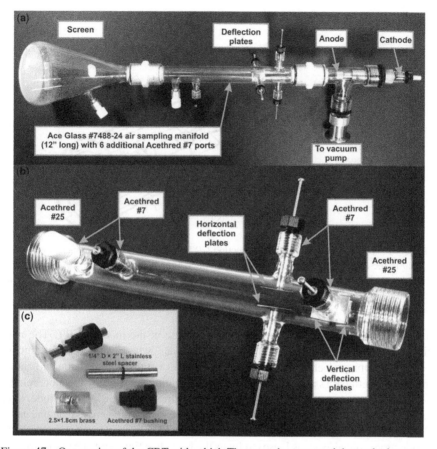

Figure 47   Our version of the CRT with which Thomson demonstrated that cathode rays are deflected by an electrostatic field. (a) We used our homemade CRT of Figure 40 and added an Ace-Thred #25 12-in. column (Ace Glass #7488-24 air-sampling manifold) that was custom-ordered to have six Ace-Thred #7 ports. (b) The deflection plates are supported by $\frac{1}{4}$-in. diameter × 2-in. long threaded spacers. (c) We cut the deflection plates from a thin sheet of brass and soldered brass screws to each.

between the plates. Do your results agree with Thomson's observation that the cathode-ray beam deflection is proportional to the potential difference between the plates?

THOMSON AND THE MODERN CRT

The tube that Thomson built for his electrostatic deflection experiment (Figure 46) is the grandfather of the modern oscilloscope CRT.** As shown in Figure 48, the basic

---

**However, German physicist Karl Ferdinand Braun developed the first CRT with magnetic beam deflection and a mica screen covered with phosphor to produce a visible spot. Braun used this tube as an indicator tube to visualize alternating currents, making Braun's invention the first oscilloscope.

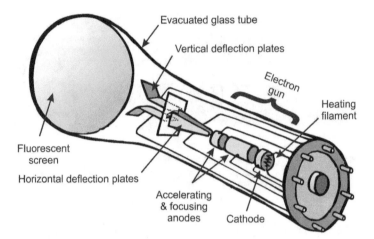

Figure 48   The modern electrostatic-deflection CRT used in CRT oscilloscopes is very similar to the one Thomson used in his electron-deflection experiment. The most dramatic change is in the use of a thermionic cathode and an additional set of deflector plates that allow the beam to be steered in both axes (horizontal and vertical).

structures of Thomson's tube are still recognizable in the modern CRT. The main difference is in the replacement of the "cold cathode," where cathode-ray production relies on stripping electrons by ionizing traces of air between the cathode and anode. Instead, modern CRTs are sealed at a very high vacuum, and employ the more reliable *thermionic emission* method by which electrons are "boiled-off" the cathode when heated by a filament. In an improvement beyond Thomson's tube, more than one anode electrode is used as part of the CRT's "electron gun." These anodes not only accelerate the electrons, but also act as lenses to focus the beam to a tight spot. The modern CRT also adds a set of deflector plates over Thomson's tube, allowing the beam to be steered vertically and horizontally to reach any point on the fluorescent screen.

Now that microprocessor-controlled LCD screens have virtually replaced CRTs, small oscilloscope CRTs are plentiful and inexpensive in the surplus market, giving us the opportunity of using them to reproduce many of the classical experiments on cathode rays without having to pay careful attention to regulating the vacuum inside the tube.[††] For our experiments, we built a rudimentary oscilloscope based on a design by J. B. Calvert of the University of Denver.[12] We used the widely available 2AP1 CRT. However, the circuit should work equally well for other CRTs with 2-in.-diameter screen, such as the 2BP1 or 902. As shown in the schematic diagram of Figure 49, we use a dual-secondary transformer to power the CRT. The low-voltage secondary heats the filament, while the current from the high-voltage secondary is rectified and fed to a resistor divider that produces the various voltages needed to operate the CRT. Although the transformer is rated at only 500 V, the loading in our circuit is very low,

---

[††]However, you may want to try a closer reproduction of Thomson's experiments once you gain experience with your vacuum system. Steve Hansen published a paper on the construction of a functional Thomson e/m apparatus. This is a great reference should you decide to undertake such a project.[13]

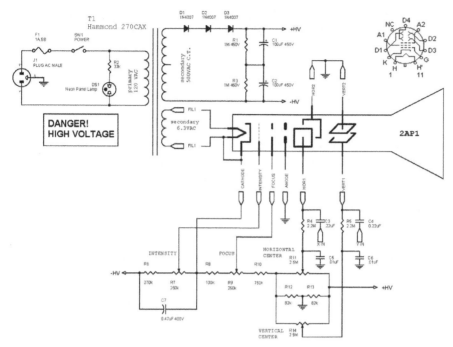

Figure 49 We built this simple oscilloscope to experiment with the properties of cathode rays. The CRT is a surplus 2AP1 tube, but you can substitute one of the many other oscilloscope CRTs with 2-in.-diameter screens. We recommend that you use banana connectors between the CRT and your circuit so that you can reuse the CRT for other experiments.

so the voltage between +HV and -HV in our oscilloscope reaches well over 700 V. Be very careful when you build this circuit, since these voltages are dangerous!

Inside the CRT, electrons are produced by a thermionic cathode when sufficiently heated by the filament. Right after the cathode, electrons are squirted out of a small hole in a cup-shaped electrode called the "intensity grid." The electrons are then accelerated and focused by two other anodes before reaching the region where they can be deflected by the deflection plates.

Potentiometers R11 and R12 place a voltage across their corresponding deflection plates. Simply turning the knobs on the horizontal or vertical centering electronics reproduces Thomson's second experiment! Bring a magnet close to the CRT and see how the beam is deflected. Pay attention to the orientation of the magnet needed to cause the same deflection of the beam as an electrostatic field.[‡‡] Another interesting experiment is to observe the deflection caused by turning the CRT around so that the electrons are affected by the earth's magnetic field.

[‡‡]To make your circuit behave as an oscilloscope, all you have to do is to inject AC signals between X_IN/Y_IN and ground. You will need approximately 150 V peak-to-peak to cause the beam to move all the way from one side of the screen to the other. You could use the output of a 60-VAC transformer to draw lines and circles. Calvert[12] has presented a simple deflection amplifier to make the oscilloscope more useful.

THOMSON'S THIRD EXPERIMENT—MASS-TO-CHARGE RATIO OF THE ELECTRON

In his third experiment, Thomson measured the mass-to-charge ratio (m/e) of the cathode rays. His setup allowed him to measure how much energy the rays carried and how much they were deflected by a magnetic field while also being deflected by an electric field.

Figure 50 shows a simplified version of the apparatus. A potential difference V accelerates electrons from the cathode toward the anode. Electrons then emerge out of the anode plate as a narrow beam. There are two large, parallel aluminum plates that are energized by a high-voltage supply to maintain uniform electric field E across a fluorescent screen overlaid with a reticule. A magnetic field is applied perpendicular to the plane of the screen. The electric and magnetic fields are directed at right angles to each other and also at right angles to the beam of electrons. For his experiment, Thomson adjusted the strengths of the two fields so that the upward deflection of the beam of electrons due to the electric field was completely cancelled by the downward deflection by the magnetic field. The charge of the electron e was unknown to Thomson at the time, so he estimated the electron's velocity by measuring the deflection of the beam with only the magnetic field on, and once again with only the electric field on. Thomson reasoned:

Force on electron due to electric field $= e \times E$

Force on electron due to magnetic field $= e \times B \times$ (velocity of electron)

$$\frac{\text{Magnetic deflection force}}{\text{Electrostatic deflection force}} = \frac{eB \times (\text{velocity of electron})}{eE}$$

$$= \frac{B}{E} \times (\text{velocity of electron})$$

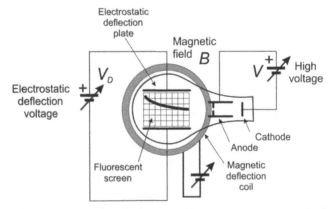

Figure 50   For his third crucial experiment of 1897, Thomson measured the mass-to-charge ratio of the electron by using a magnetic field to balance the electrostatic deflection of a cathode-ray beam.

where the magnetic deflection in Thomson's experiment is set to exactly balance the electrostatic deflection, so:

$$1 = \frac{B}{E} \times \text{(velocity of electron)}$$

At the same time, the electron's velocity may be found from the energy-balance equation for the kinetic energy acquired by the undeflected electron when accelerated by the accelerating voltage V:

$$\text{Kinetic energy of electron} = \frac{1}{2}m\,\text{(velocity of electron)}^2 = eV$$

Replacing the velocity of the electron from the prior equations, and rearranging the terms to measure e/m:

$$\frac{e}{m} = \frac{\text{(velocity of electron)}^2}{2V} = \frac{E^2}{2VB^2}$$

Thomson estimated m/e to be between 1.1 to 1.5×10^{-11} kg/Coulomb using his crude CRT. The modern value is 0.57×10^{-11} kg/Coulomb, so he did a very good job at estimating m/e's order of magnitude, which allowed him to state that the mass-to-charge ratio of the new particle was over 1,000 times lower than that of the lightest atom known—a hydrogen ion (H^+)—suggesting either that the particles were very light and/or very highly charged.

Thomson thus concluded:

> As the cathode rays carry a charge of negative electricity, are deflected by an electrostatic force as if they were negatively electrified, and are acted on by a magnetic force in just the way in which this force would act on a negatively electrified body moving along the path of these rays, I can see no escape from the conclusion that they are charges of negative electricity carried by particles of matter.

In the years that followed, Thomson figured out ways of directly measuring e, allowing him to determine the mass of the negatively charged particles in cathode rays. The measurements showed that e is the same in magnitude as the charge carried by the hydrogen ion when water is split up by electrolysis. Thomson proved that the mass of the particles was less than 0.001 of the mass of the hydrogen atom—the smallest mass known at that time. Thomson believed the atomic diameter to be around 10^{-10} m, which is well within the modern range between 60 and 600×10^{-12} m. He also estimated the electron's size to be approximately 10^{-13} m, which compares well to the so-called "classical electron radius" equal to 2.82×10^{-15} m.

Thomson called the particles that he discovered *corpuscles*.[14] Today, we call these particles *electrons* after the "fundamental unit quantity of electricity" proposed by Irish physicist George Johnstone Stoney in 1891.

MEASURING e/m WITH OUR CRT

It is possible to use a modern CRT to measure e/m using Thomson's method. However, it isn't an easy task, because the deflection plates are relatively small, and their position is not well known due to the graphite coating used to reduce charge accumulation beyond the neck of the CRT. For this reason, specialized CRTs are made by didactic equipment companies to enable students to reproduce Thomson's third experiment. One such tube sometimes found in the surplus market for around $150 is the model TEL-525 made by Tel Atomic.[§§]

However, the same inexpensive 2AP1 CRT that we used in our oscilloscope can be used to measure e/m using a modification of Thomson's method developed by J. B. Hoag[15,16] in the 1920s. As shown in Figure 51, the idea is to place the CRT within a solenoid while connecting an AC voltage across one set of deflector plates. The AC signal produces a line on the face of the CRT. As the current increases through the solenoid, this line appears to rotate and shorten because of an effect known as *magnetic focusing*. An analysis of this effect is beyond the scope of this book, but can be found in Hoag's original book, as well as in reference 17.

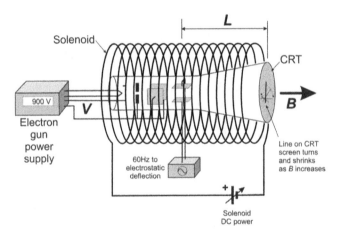

Figure 51   This simplified diagram shows how you can use your CRT to measure e/m using Hoag's method. An AC signal is placed across one set of plates of the CRT to produce a line on the screen. The solenoid is then energized until the line makes one complete helical turn. You can calculate $e/m = 8\pi^2 V/B^2 L^2$ using the accelerating voltage V between the cathode and the anode, as well as the magnetic field B needed to cause one complete turn of the line on the screen until it disappears into a small point. For the sake of clarity, we show a short solenoid. However, the solenoid should extend well past both ends of the CRT to ensure a homogeneous magnetic field.

[§§]The TEL 525 deflection tube and matching TEL 502 Helmholtz coils have been produced by Tel Atomic for over 30 years, and are still available directly from them, although the deflection tube carries a hefty price for an amateur experimenter.

At some point, the line would have made a whole turn and shrunk to a point. Just a small spot will appear on the screen. *e/m* can then be calculated from the magnetic field *B* needed to cause the line to rotate one full turn until it disappears into a point:

$$\frac{e}{m} = \frac{8\pi^2 V}{B^2 L^2}$$

where *V* is the accelerating potential (voltage between anode and cathode of CRT) and *L* is the length of the helical path (distance between deflector plates and CRT screen).

The actual circuit for our setup is shown in Figure 52. We gutted a surplus power supply that we bought from Surplus Sales of Nebraska and used its original transformer and chassis. The specific transformer is not critical. Just use any transformer with ~150 VAC and 6.3 VAC secondaries. The only thing that you must check is that the transformer's insulation is rated at 1,000 V or higher. Another possibility is to use a 6.3 VAC "filament" transformer and a separate 110 VAC/110 VAC isolation transformer (e.g., Magnetek N-48X). The over-100 VAC transformer is used to oscillate the electrostatic deflection plates, so selecting a specific voltage is not important.

The accelerating voltage is obtained from an external high-voltage power supply. Either one of the high-voltage power supplies that we built in chapter 1 (Figure 31 or Figure 32) works well. This power supply is not connected directly to the cathode (because the intensity grid electrode must be more negative than the cathode), so a digital multimeter must be connected between the cathode and the anode to measure the accelerating potential *V*.

The solenoid coil is made of 16 AWG enameled magnet wire. Ideally, the coil should extend past the ends of the CRT by at least 5 cm, so you could wind a coil that is at least 30-cm-long. A 34-cm-long piece of 2-in. ID PVC pipe is a good center form for the solenoid. Wind five smooth layers of the magnet wire, making sure that you count the total number of turns that you wind on the solenoid. Our solenoid has over 1,100 turns.

We have a Gaussmeter available (an instrument to measure magnetic field), so we were able to find out quite precisely that the field produced by our solenoid is given by $B[\text{Tesla}] = 0.004187 \times \text{Current}[A]$. However, the reason why we asked you to count the number of turns on your solenoid is because you can calculate the magnetic field *B* inside your solenoid as you energize it with current *I* if you know the total number of *turns* in your coil and the length L_{solenoid} they occupy:

$$B[\text{Tesla}] = 4\pi \times 10^{-7} \left[\frac{V \cdot s}{A \cdot m}\right] \times I[A] \times \frac{turns}{L_{\text{solenoid}}[m]}$$

To conduct an experiment, you would set the accelerating voltage of the tube to an initial value $V = 300$ V. Set the controls to produce a thin vertical line. Start increasing the current through the solenoid until a full turn is traced by the line on

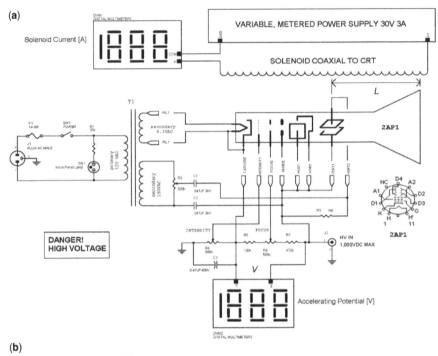

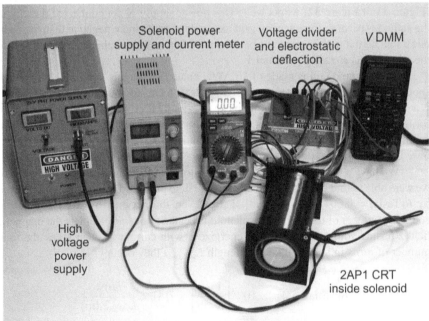

Figure 52   We built this setup to measure e/m using Hoag's method. (a) Schematic diagram of the circuit to energize the CRT and coaxial solenoid. High voltage is obtained from an external power supply. (b) A view of our setup ready to perform an experiment.

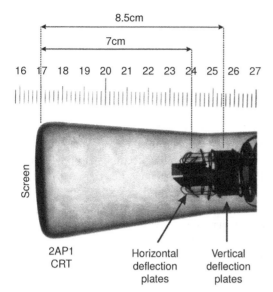

Figure 53   X-ray of the 2AP1 CRT showing the distances from the screen to the middle of the deflection plate pairs.

the CRT. Record the accelerating potential and current through the solenoid. Repeat at various accelerating potentials (e.g., taking steps of 100 V up to a maximum of 800 V).

You will need to know the length L of the helical path. This is the distance between the center of the deflection plates and the screen. However, the graphite coating inside the tube will prevent you from measuring this directly, so we took an X-ray of the 2AP1 tube and found out these distances to be approximately 8.5 cm for one set of deflection plates, and 7 cm for the other set of plates (Figure 53).[¶¶] You may use the same value if you are using a 2AP1 CRT.

Finally, calculate $e/m = 8\pi^2 V/B^2 L^2$ for each run and yield an average estimate for e/m. How well do your measurements compare with the accepted value of $e/m = 1.76 \times 10^{11}$ Coulomb/kg?

A MAGICAL MEASUREMENT OF e/m

An even simpler, although less accurate, adaptation of Thomson's setup to measure e/m can be built using a surplus "magic eye" tube. These electron vacuum tubes were commonly used in tube radios as a visual aid for tuning. The purpose of a magic eye tuning tube in these radio receivers was to help tune a station at its strongest point on the dial. The visual aid of the tube made variations in signal strength more obvious than by ear alone.

As shown in Figure 54b, the 6AF6-G tube that we use has a cylindrical cathode heated from its inside by a wire filament. Electrons emitted from the cathode are

[¶¶]Which set of plates is "vertical" and which is "horizontal" depends only on the orientation of the tube.

accelerated horizontally by a high voltage (commonly in the ballpark of 150 V) toward a bowl-shaped anode. As the electrons strike the anode, they cause the phosphor coating on the inside of the bowl to glow. Two "ray-control" electrodes control the size of wedge-shaped shadows produced on the anode's fluorescent coating. A positive voltage applied to a control electrode reduces its corresponding shadow.

To measure e/m, we placed a large coil of wire (an air-core solenoid) around the magic eye tube. A current passing through the solenoid subjects the electrons to a vertical magnetic field, deflecting their paths. Without the field, we can assume that the electrons move horizontally at a constant velocity determined by the voltage between the cathode and the anode. When the field is applied, the electrons will be deflected into a circular path, as shown in Figure 54c. We can thus determine e/m if we know the strength of the magnetic field and the radius of the circular path.

The setup is quite simple. We attached a tube socket to a small plastic enclosure, and wired the socket's contacts to female banana connectors. Next, we cut a 4-in.-long piece off a 2-in.-diameter plastic pipe used in plumbing for sink drains. We used 28AWG enameled magnet wire to wind three layers of turns on the plastic tube. We placed the resulting solenoid on the tube. We wired the tube and solenoid to three power supplies, as shown in Figure 54a. Either one of the high-voltage power supplies that we built in chapter 1 (Figure 31 or Figure 32) works well, but one must be very careful not to exceed the 6AF6's 200-V limit.

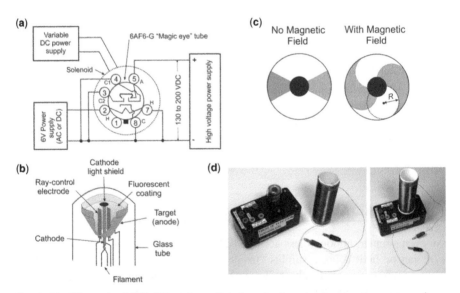

Figure 54   We used a 6AF6-G "magic eye" tuning tube in a simple setup to measure e/m. **(a)** Circuit diagram of our setup. **(b)** In the 6AF6, electrons produced by a thermionic cathode cause fluorescence on the tube's anode. **(c)** Applying an external magnetic field curves the path of the electrons reaching the anode's fluorescent coating. Knowing R and the voltage applied to the tube allows one to calculate e/m. **(d)** We mounted a socket for the tube to a small plastic box, and used banana connectors to make the setup easy to use. The solenoid's core is a 4-in. section of 2-in. plastic strainer tailpiece pipe.

Build one yourself and try it out! None of the component values are critical, so feel free to substitute to accommodate what you have in hand. 6AF6-G tubes go for around $20 on eBay. If you don't have an octal tube socket, or prefer not to solder to one, you may use a relay octal socket base with screw terminals as a very convenient socket for the tube. Oh yeah—make sure you count the exact number of turns that you wind on your solenoid.

We used our Gaussmeter to find that the field produced by our solenoid in the region of the tube's anode is given by $B[\text{Tesla}] = 0.0079 \times \text{Current}[A]$. However, as we did before, you could also calculate the magnetic field B inside your solenoid as you energize it with current I if you know the total number of turns in your coil and the length L they occupy:

$$B[\text{Tesla}] = 4\pi \times 10^{-7} \left[\frac{V \cdot s}{A \cdot m} \right] \times I[A] \times \frac{\text{turns}}{L_{\text{solenoid}}[m]}$$

To perform the experiment, dim the lights in the room, turn on the filament, power the tube with 130 V, and wait until it glows. Use a set of cylindrical wooden dowels of known diameters (e.g., $\frac{1}{4}$, $\frac{1}{2}$, and $\frac{3}{4}$ in.), and increase the current through the solenoid until the curve traced by the electrons matches the curve of the dowel. Repeat the measurements at tube voltages V between 90 and 190 V (make sure that you don't exceed the 6AF6's 200-V maximum!)

Use your data to estimate the mass of the electron using the formula:

$$m_e = \frac{B^2 e R^2}{2V}$$

where $e = 1.60 \times 10^{-19}$ Coulomb. How well does your estimate match the accepted value of $m_e = 9.11 \times 10^{-31}$ kg?

THOMSON'S "PLUM PUDDING" MODEL OF THE ATOM

Thomson concluded that the negatively charged particle of cathode rays must be a fundamental part of matter itself. His model presented the atom containing a large number of smaller bodies, which he still called *corpuscles*. Since common atoms are electrically neutral, Thomson proposed that the atom comprises separate negative and positive parts. The negative corpuscles (electrons) were the carriers of negative electrical charge, and the positive ion left behind was a bubble with much larger mass than the negative corpuscles.

At a time before the discovery of the atomic nucleus, Thomson imagined that "the atoms of the elements consist of a number of negatively electrified corpuscles enclosed in a sphere of uniform positive electrification."

Thomson's 1904 model was compared by his fellow scientists to a British dessert called plum pudding (Figure 55): the atom is composed of electrons surrounded by

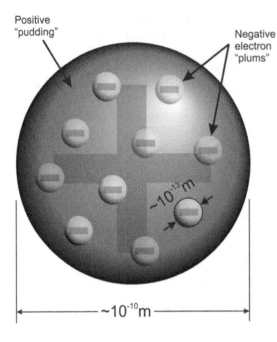

Positive
"pudding"

Negative
electron
"plums"

~10^{-13}m

~10^{-10}m

Figure 55   In Thomson's "plum pudding" model of the atom, tiny electron "plums" float inside a much larger blob of positively charged "pudding." Ernest Rutherford proved this model wrong in 1911 after he discovered the atomic nucleus.

a soup of positive charge to balance the electrons' negative charges, like negatively charged "plums" surrounded by positively charged "pudding."

This model was shown to be false by an experiment conducted by Ernest Rutherford in 1911, in which Rutherford discovered that the small nucleus of the atom contained a very high positive charge; this discovery led to the Rutherford model of the atom.

GEIGER–MÜLLER COUNTER

We will need a radiation counter to continue our exploration of the subatomic world, so let's discuss gas-filled radiation detectors, especially the type commonly known as a Geiger counter.

As shown in Figure 56, a gas-filled radiation detector is simply a metal cylinder filled with an inert gas. A thin center wire is kept at high potential, usually between 300 and 3,000 V. The gas is ionized whenever a high-energy particle or a gamma ray enters the cylinder, causing a current pulse that can be detected after it passes through an amplifier. German physicists Hans Geiger and Walther Müller developed a tube in which radiation strips electrons off the neutral gas atoms, yielding positively charged ions and electrons. Both of these accelerate toward their oppositely charged electrodes. As they move, the ion pairs gain sufficient energy to ionize further gas molecules through collisions on the way, creating an avalanche of charged particles. This causes an intense pulse of current that cascades from the negative electrode to the positive electrode. This pulse is easily detected by very simple electronics.

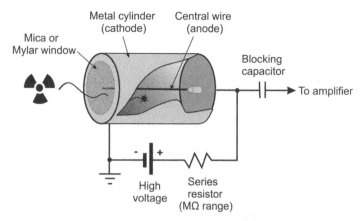

Figure 56   When ionizing radiation enters a gas-filled detector, the gas inside the cylinder is ionized, causing a pulse of current between the anode and cathode. In a GM tube, the gas mixture and operating voltage are chosen so that a single ionizing event causes a cascade of ionization to produce a more easily detectable pulse.

The material used for the window determines the types of radiation that can enter the tube to be detected. Tubes with a glass window will not detect alpha radiation, since it is unable to penetrate the glass, so they are limited to detecting beta radiation and gamma rays (including X-rays). Tubes with a mica or Mylar window will detect alpha radiation, but the window area is very fragile. Geiger–Müller tubes don't usually detect neutrons, since these do not ionize the gas inside the cylinder. We will discuss the different types of radiation later in this chapter.

The design of a Geiger counter is relatively simple and straightforward. In addition, complete units are widely available in the surplus market. These counters were distributed to fallout shelters by the Civil Defense in the 1950s and 1960s for use by civilian survivors in case of nuclear war. However, most have been taken out of service since the end of the Cold War, making new units plentiful.

Radiation-survey enthusiast George Dowell (owner of GeoElectronics) started the trend of modifying the Electro-Neutronics (ENi) Civil Defense V-700 model 6b into a clone of the same type of unit manufactured by Lionel.[18] The version made by Lionel has a much better circuit, but is much more difficult to find. The ENi model has a great mechanical layout, so the modifications are easy to make.

We took the concept further and made a number of changes, improvements, and addition of features that turn a stock ENi CD V-700 into a great instrument for lab use (Figure 57). We affectionately call our version the "CD V-700 Pro." Our version is based on George's "LENi" modification, but adds the following features:

- Preamplifier to make it compatible with photomultiplier scintillation probes (GM tubes can still be used).
- Selectable, regulated bias voltage (900 or 1,200 V) for connection to GM tubes and PMTs. A blinking indicator warns of the high-voltage selection.
- Noise-reduction circuitry eliminates hum.

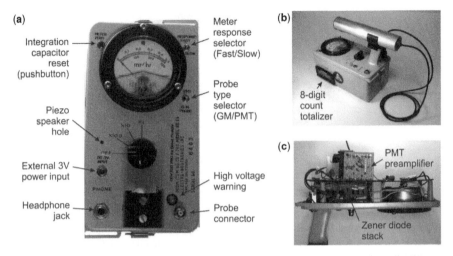

Figure 57 We modified a surplus Civil Defense V-700 radiation survey meter made by Electro Neutronics (model 6-b) into a very capable radiation counter capable of working with both GM and PMT scintillation probes. (a) We modified the front panel to accommodate the new switches, connectors, and panel light. (b) We placed a Veeder-Root count totalizer module on the side of the box. (c) The new electronic components, including a Zener diode stack and a PMT preamplifier are wired directly to the original printed circuit board.

- Internal piezo clicker.
- Power input jack saves batteries when powered from car or AC-operated power supply.
- 8-digit digital counter.

As a first step, study the schematic diagram of the original ENi CD V-700 (Figure 58) and identify the components on the PCB. Then study the schematic diagram of the instrument with the "LENi" modification (Figure 59) and final version (Figure 60). Start the modification by disconnecting the GM probe from the circuit. Then dig into the circuit and modify the power supply section:

- Remove Zener diode CR6 (sometimes 2 diodes in series) and discard, it will not be used.
- Remove R13 and reinstall it in series with the base lead going to the oscillator transistor.
- Add a 0.0022-μF, 50-V capacitor between the base and collector of transistor V4 (Q4).
- Substitute CR5 (D5) by a modern >5-kV silicon diode. We used a Fuji ESJA53-20-A, 20-kV, 5-mA diode.
- Substitute C8 with a modern 0.01-μF, 3-kV capacitor.
- Substitute C5 with a modern 0.0022-μF, 3-kV capacitor.
- Replace R13 by a 1.8-MΩ resistor and a 3.3-MΩ resistor in series.

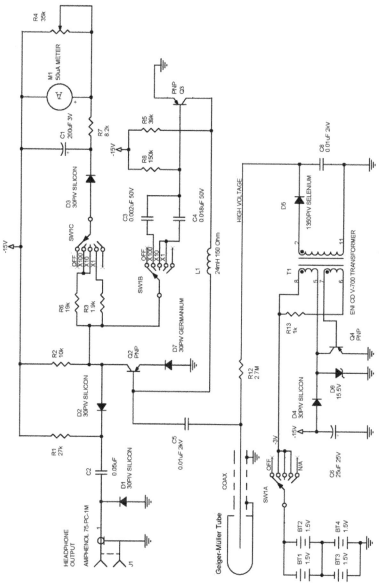

Figure 58   This is the circuit schematic for the V-700 model 6-b Geiger–Müller Survey Meter made by Electro-Neutronics (ENi).

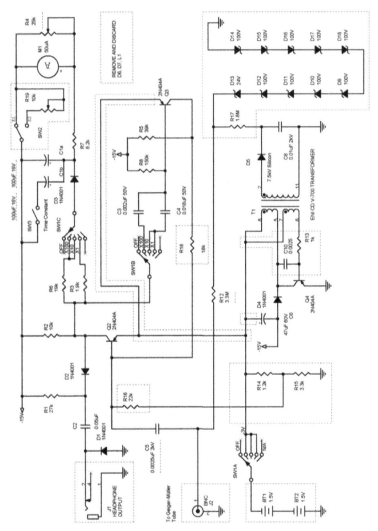

Figure 59 The ENi CD V-700 model 6-b can be modified to improve its performance, which turn it into a great lab instrument at a bargain price. The LENi modification by George Dowell is the most popular adaptation among radiation-survey enthusiasts.

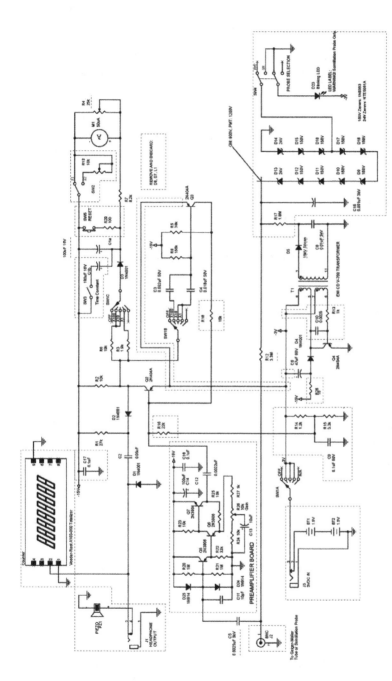

Figure 60 If you are already modifying an ENi CD V-700 model 6-b, you may want to follow our circuit. Our version adds a piezo clicker, a digital totalizer, and compatibility with scintillation probes based on PMTs. We affectionately call this version the "CDV-700 Pro."

At this point you will need to build a specialized $\times 10$ high-voltage probe by stringing nine 10-MΩ resistors in series. An empty ballpoint pen makes a handy enclosure for the probe. Connect the probe to a digital multimeter with 10-MΩ input impedance when set to a range suitable for measuring 200 VDC (use a second multimeter to measure the resistance across the input terminals of your DMM), and measure the voltage across C8. With no GM tube connected, you should measure between 1,300 and 1,900 VDC (130 to 190 V on your multimeter).

Next, build the Zener diode stack on a small piece of perforated board. Build the Zener diode stack using 9 1N5383 150 V Zeners and 2 NTE5081A 24 V Zeners. Wrap the stack with a layer of Kapton™ tape, and then add C15 to the back side of the board. Wrap the assembly with Kapton tape. Mount the Zener diode regulator board onto the CD V-700 PCB. Check your progress: if you power the instrument, the voltage across C15 should now be around 1,200 V, and approximately 900 V if you short the cathode of D17 to ground.

Now add the biasing circuit comprising R14, R15, R16, and C9. Continue by modifying the metering circuit:

- Remove CR7 (D7). Connect the emitters of the metering transistors in parallel and route them to the -3 V line.
- Replace L1 with an 18-kΩ, $\frac{1}{2}$-W resistor (R18 in the CDV700 Pro schematic).
- Modify the circuit to insert a 10-Ω, $\frac{1}{2}$-W resistor between the anode of D4 and the "-15 V" line feeding the metering circuit.
- Replace C6 with a 47-μF, 60-V electrolytic capacitor.
- Replace C1 with two 100-μF capacitors. Leave the negative terminal of one of these capacitors open so that it can be connected to the front-panel, time-constant selection switch.

Next, you will need to modify the CD V-700's front panel. Start by installing a BNC connector for the probe. Remove and discard the sealing nut through which the GM probe cable passed. Tap this hole using a $\frac{3}{8}$-in.-diameter 32 tpi pitch tap (McMaster-Carr 25705A64). Mount a nonisolated bulkhead BNC connector (e.g., Jameco 71589) on this tapped hole. Drill the front panel (use casting marks on the back side) to accommodate the extra switches, connectors, piezo speaker, and LED, as shown in Figure 57a. Use a nibbling tool to cut a 68-mm $\times$ 33-mm rectangular hole on the enclosure bottom to accommodate the Veeder-Root A103-000 totalizer, as shown in Figure 57b.

Once you are done with the mechanical changes, wire the instrument as shown in Figure 60, using clean, new cable with insulation for the appropriate voltage rating. Route cables next to the enclosure and keep connections as short, direct, and clean as possible. Use high-voltage test lead wire between the PCB and the center terminal of the probe BNC. Use good heat-shrink tubing to dress all switch connections. Keeping things tidy will really pay off later. Once the PMT preamplifier is added, noise will creep into the system if you don't pay attention to your wiring habits.

This is a good point to test your "almost LENi". The voltage at the BNC connector measured with the high-impedance $\times 10$ probe should read around 1,200 V with the probe selector switch in the "Scintillator" position. The flashing LED should light up and blink. Flipping the probe selector switch to the GM position should cause the LED to turn off, and the voltage at the connector should drop to approximately 900 V.

Turn the probe selector switch to the "GM" position. Connect a GM probe (e.g., the original CD V-700 probe to which you have installed a male BNC connector). The unit should produce background clicks and be able to detect the radiation emitted by the operational check source under the CD V-700's original label. Next, connect the totalizer module to the circuit and verify that the counter advances once for every "click."

Build the preamplifier circuit on a small piece of prototyping board. Keep wires short and the circuit neat and organized. Unsolder C5. Mount the preamplifier board directly onto the CD V-700 PCB, as shown in Figure 57c. Add the ground wire and bypass capacitors C16 and C17.

Add a two-D cell plastic battery holder to make space for the preamplifier circuit. Drill some mounting holes on the PCB and mount the battery holder using $\frac{1}{4}$-in. nylon spacers. CONGRATULATIONS! You have completed modifying your surplus ENi CD V-700 model 6b into a CDV-700 Pro!

The CDV-700 Pro is compatible with virtually any GM tube that operates on 900 V. We recommend that you purchase a tube with a mica window (sensitive to alpha radiation) besides the probe that came with your CD V-700, which is sensitive only to beta- and gamma-rays. A suitable tube is the LND 7311 "pancake" GM tube, or a probe made with this tube, such as the GeoElectronics GEO-210.

You can also use the CDV-700 Pro with sensitive scintillation probes. *Scintillators* are materials that produce a short flash of light when hit by ionizing radiation. We already met one such scintillator when we used zinc-selenide as a phosphor that emits light when struck by electrons inside a CRT. Other purpose-grown inorganic crystals have been developed and are commonly used in physics labs. The most widely used is NaI(Tl) (sodium iodide doped with thallium). In addition, many organic liquids and solids exhibit scintillation. Easy-to-use plastic scintillators have been developed by incorporating organic scintillating substances within a transparent plastic. Plastic scintillators are widely used by experimenters, because they are relatively inexpensive and are very easily shaped.

Detection of the scintillation from a scintillator crystal or plastic is commonly done with a PMT. For example, a cylindrical piece of plastic scintillator can be coupled to the PMT probe we built in chapter 2 (Figure 30). However, for the CDV-700 Pro, the PMT has to be wired in a slightly different way than we did in Figure 29. Figure 61 shows the schematic diagram for a scintillation probe powered directly by the CDV-700 Pro. Note that the resistors are of much larger value than those we had used for the probe of Figure 29. This is because the CDV-700 Pro's power supply cannot provide more than just a couple of microamperes. We used a spare Photonis XP2102, but almost any 2- to 3-in. PMT should work equally well if you follow the general circuit and adapt it to the pinout and number of dynodes of your PMT.

The scintillator should be coupled to the PMT's face with index-coupling grease. This doesn't have to be the expensive gel sold for premium optical systems (e.g., Dow Corning Q2-3067 optical coupling compound at over $250 for a 4-oz jar), but rather any low-cost, high-purity silicone grease such as Dow Corning 4. Use a tiny bead and let it squeeze between the PMT and the scintillator to form the

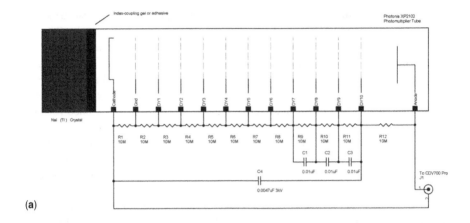

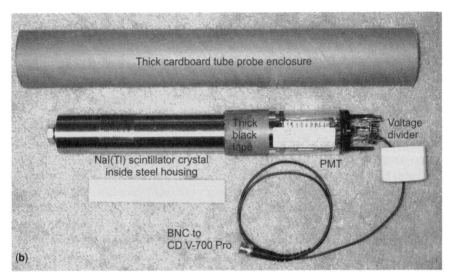

Figure 61 A crystal or plastic scintillator coupled to a PMT can be used as a very sensitive radiation probe for the CDV-700 Pro. (a) Our probe is self-contained, and is powered directly via the CDV-700 Pro's probe connector. (b) The complete assembly must be built inside a light-proof enclosure. We used a thick cardboard shipping tube and matching tin caps to enclose the probe.

thinnest possible interface between the two components. If you prefer not to use index-coupling grease, you may omit it, but this omission will result in a 10–20% drop in probe sensitivity.

Lastly, package the complete assembly within a light-tight enclosure. We used a cardboard shipping tube with 2-in. ID and its matching tin caps to shield the PMT/scintillator assembly from light, but any other light-tight enclosure will work well. Radiation-survey aficionado Charlie Thompson built a ground-survey probe by cleverly enclosing a PMT coupled to a square block of plastic scintillator inside a Home Depot® paint can.[19]

α, β, AND γ

Thomson's discovery of the electron demonstrated that the phenomenon of ionizing radiation was not a single process. However, the connection between Thomson's electrons, Roentgen's X-rays, and the Becquerel "uranium radiation" was a complete mystery.

In 1898, Pierre and Marie Curie discovered that thorium (a metal represented by the symbol Th and atomic number 90) gives off "uranium rays," which Marie renamed *radioactivity*. Later that year, the Curies discovered the elements polonium and radium. The Curies realized that radioactivity must be some new property of the atoms themselves, since the long series of chemical processes used to isolate them from radioactive rocks didn't change their radioactive emissions.

British/New Zealander physicist Ernest Rutherford—then a research student at the Cavendish Laboratory—showed that X-rays and radioactivity were capable of ionizing gases in essentially the same way. A year later, Rutherford noticed that uranium salts produced two different types of radiation—one, which he termed alpha (α) radiation, was highly ionizing but unable to penetrate even thin pieces of paper; the other type of radiation, which he termed beta (β) radiation, exhibited lower ionizing capability but was of higher penetrating power. In 1903, Rutherford realized that a type of radiation from radium discovered (but not named) by French chemist Paul Villard in 1900 must be different from alpha rays and beta rays, due to its much greater penetrating power. Rutherford named it gamma (γ) radiation (Figure 62).

Let's evaluate the penetrating power of alpha, beta, and gamma radiation for our first experiment with the CDV-700 Pro Geiger counter. You will need a GM tube such as the LND 7311 "pancake" that is sensitive to all three types of radiation. If you prefer, you could purchase a ready-made probe made with this tube, such as the GeoElectronics GEO-210. Please note that the mica window in these GM tubes is extremely fragile, so be very careful to avoid puncturing it and ruining your detector!

Next, you will need three radioactive sources, each with a dominant emission. Table 4 and Figure 63 show some of the radioactive sources available to the

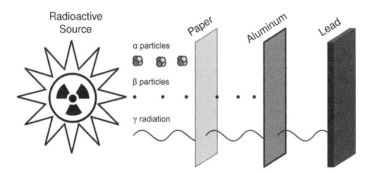

Figure 62   Ernest Rutherford classified radiation based on its penetrating power. He called the radiation that is highly ionizing, but which cannot penetrate a sheet of paper, *alpha* (α). He called radiation that had lower ionizing power, but higher penetrating power *beta* (β). Rutherford named the most highly penetrating rays *gamma* (γ) radiation.

TABLE 4 Some Radioactive Sources that may Be Used for Experimenting with the Penetrating Power of Alpha, Beta, and Gamma Rays*

Radioactive material	Main emission	Half-life	Sources
Polonium-210 (^{210}Po)	α	138.4 d	• 500-μCi Staticmaster® static-eliminator cartridges • 0.1-μCi exempt, solid-disk source
Americium-241 (^{241}Am)	α	432.2 yr	• 1-μCi source from ionization-type smoke detector
Thorium-232 (^{232}Th)	α	1.4×10^{10} yr	• Old gas lamp mantles • Thoriated tungsten welding electrodes
Radium-226 (^{226}Ra)	α	1,601 yr	• Old luminescent watch/clock hands • Old glow-in-the-dark aircraft switches and dials • Westinghouse 1B45 radar tube
Uranium-238 (^{238}U)	α	4.5×10^{9} yr	• Uranium-glass marbles • DU bullets • Old Fiesta® ceramic dinnerware
Strontium-90 (^{90}Sr)	β	28.5 yr	• 0.1-μCi exempt, solid-disk source
Carbon-14 (^{14}C)	β	5,730 yr	• 10-μCi exempt, solid-disk source • Trace amounts in every living thing, including Creationists
Tritium (^{3}H)	β	12.32 yr	• Tritium gas inside some modern luminescent signs, gun sights, and watches • 200-mCi source in 3M model 703 static meter
Barium-133 (^{133}Ba)	γ	10.7 yr	• 0.1-μCi exempt, solid-disk source
Cesium-137 (^{137}Cs)	γ	30.17 yr	• 10-μCi exempt, solid-disk source • TG-77 spark gap tube
Cobalt-60 (^{60}Co)	γ	5.27 yr	• 1-μCi exempt, solid-disk source
Europium-152 (^{152}Eu)	γ	13.5 yr	• 1-μCi exempt, solid-disk source
Technetium-99 (^{99m}Tc)	γ	6 h	• 10 to 30 mCi inside uncle Marty after a cardiac stress test with Cardiolite®

*This list includes items that may be purchased in the hardware store. However, it must be noted that it is illegal in the United States to disassemble these products (yes—even a cheap smoke detector) to remove the radioactive source. Fortunately, safe sources of known activity may be purchased from a number of suppliers. The solid-disk sources listed here are exempt from licensing, and require no special handling, storage, or disposal.

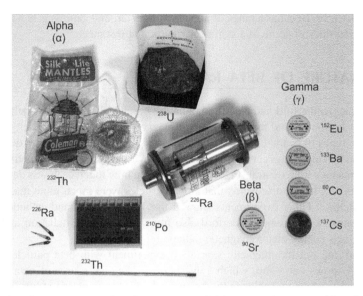

Figure 63 There are many radioactive sources available to experimenters. Although the best sources for our purpose are sealed disks that contain exempt quantities of known radioactive materials, there are everyday objects such as smoke detectors, old lantern mantles, and watches with radium-painted dials that emit detectable levels of alpha, beta, and gamma radiation.

experimenter. Safe sources of known activity may be purchased from a number of suppliers including Spectrum Techniques, Images Scientific Instruments, and United Nuclear. These are sources encased in a plastic disk or metallized onto some other substrate. They are exempt from licensing, and require no special handling, storage, or disposal. We used Polonium-210 (^{210}Po), Strontium-90 (^{90}Sr), and Cobalt-60 (^{60}Co) exempt, plastic-disk sources to experiment with the penetrating power of alpha, beta, and gamma.

   In addition to the prepackaged sources, there are a number of everyday objects such as ionization-type smoke detectors, old lantern mantles, and watches with radium-painted dials that emit detectable levels of alpha, beta, and gamma radiation. Most of these items are no longer manufactured, as exposure to the radioactive material was a health threat to the employees who made them. However, they can be found at online auctions, garage sales, and antique shops. These items were not designed for instructional or experimental use, and may therefore be hazardous when used for other purposes than originally intended. For this reason, our recommendation is to use the purpose-made, encased plastic-disk, exempt sources whenever possible.

   The experiment is simple—place each of the available sources, one at a time, close to the window of the GM tube. Pay attention to the clicking rate. Put paper, acrylic, aluminum, and lead absorbers one at a time between the source and the tube. Just a sheet of paper stops alpha radiation, a sheet of aluminum or acrylic stops beta radiation, while gamma is stopped only by a sufficiently thick sheet of lead. You can also try moving each source away from the detector. Alpha has a very short range and quickly deposits all of its energy in the air between the source and the detector. Beta has a range of about 10 cm in air, and gamma gets weaker with distance but doesn't come to a stop at any particular distance.

THE NATURE OF BETA RADIATION

At the time, Rutherford had classified radiation based only on its penetrating power. However, he soon found that an electric or magnetic field could split such emissions into three types of beams. Based on the direction in which the rays were deflected, it was found that alpha rays carried a positive charge, beta rays carried a negative charge, and gamma rays were neutral (Figure 64). Based on how much each was deflected by the same field, it was discovered that alpha particles were much more massive than beta particles.[20] Other experiments showed that beta radiation and cathode rays are both streams of electrons. Rutherford also found that gamma radiation and X-rays are both high-energy electromagnetic radiation.

Carrying out the magnetic separation experiment with beta particles is easy. Figure 65 shows our setup, which consists of a ^{90}Sr disk source of beta particles, two copper washers to collimate the beam, and a GM tube placed at 90° to the beta particle beam. A sufficiently strong magnetic field (around 800 Gauss = 0.08 Tesla) provided by a permanent magnet bends the beam so much that it may even be detected at a right angle. When you conduct this experiment be careful not to get the magnet too close to the GM tube, especially if you are using one that is also sensitive to alpha particles, since any components inside the tube that the magnet may move could damage it.

THE IONIZING POWER OF ALPHA

The experiment on the penetrating power of radiation (Figure 62) could lead you to believe that alpha radiation is weak because it is so easily absorbed. However, the

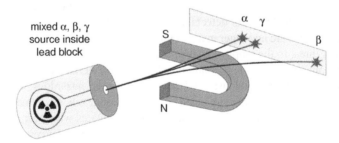

Figure 64   A magnetic field can split the radiation from a combined alpha, beta, and gamma source into three beams. Because of their positive charge, alpha rays are deflected to one side of the gamma beam, and negatively charged betas to the other side. Alpha particles are deflected much less than beta particles by the same beam, because the former are much more massive than the latter. Gamma rays are not deflected at all, because they are chargeless photons.

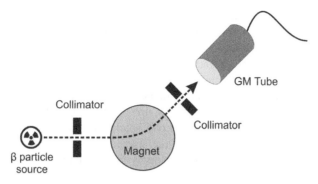

Figure 65   You can bend a collimated beam of beta particles using a strong magnet. We use a ^{90}Sr disk source of beta particles, two copper washers to collimate the beam, and a rare-earth magnet to bend the beam.

exact opposite is true. Alpha particles are so massive that they lose their energy by ripping atoms to pieces as they fly through matter. Eventually, alpha particles lose all their energy and stop harmlessly.

A simple but interesting experiment to demonstrate this point is shown in Figure 66. This is a homemade spark counter, in which a thin wire is placed at high voltage referenced to a nearby ground plane. Without a source of ionizing radiation nearby, some current will leak from the wire through the corona effect, producing a hissing noise. However, when a source of alpha radiation is brought nearby, the alpha particles passing close to the wire leave behind a trail of ions that render the air conductive, causing a spark to jump between the wire and the ground plane.

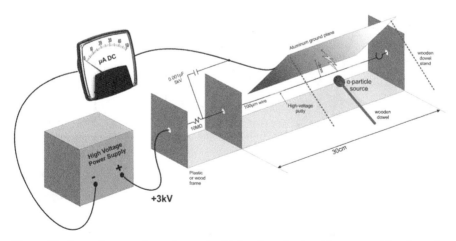

Figure 66   In our homemade spark counter, high voltage is placed between a thin wire and a large ground plane. Alpha particles ionize the air between the wire and the ground plane, causing sparks to jump between the two. Beta and gamma radiation from sources with the same activity do not ionize the air sufficiently to cause sparking.

Our spark counter follows a design by Klein et al.[21] The ground plane is a sheet of aluminum 25-cm wide × 10-cm tall. We used thin, bare copper wire for the center electrode. The resistor and capacitor are rated at 5 kV. We found that placing the aluminum sheet on an incline at a distance of around 2 to 3 mm away from the wire causes a steady corona current of around 10 μA, which we measure using a cheap analog multimeter. We had to dampen vibrations on the wire with a few beads of putty to prevent sparks caused by alpha particles from triggering continuous sparking due to shaking of the wire. Loud sparks are produced when we bring a source of alpha particles (e.g., StaticMaster cartridge containing [210]Po) close to the wire. We used the high-voltage power supply that we built to energize our CRTs (Figure 39).

Go ahead—build one yourself and try it out with different sources of radiation. Just remember to exercise care! The setup uses high voltage, and will retain some charge even after you turn the power supply off. A 3,000-V charge produces a nasty jolt, so make sure that you use a dry wooden dowel or plastic rod to bring your radioactive sources close to the charged wire. Look back at the diagram of the GM tube of Figure 56. Do you see a similarity to the spark counter you just built?

As an interesting aside, ionization-type smoke detectors take advantage of the ionizing power of alpha particles by using a steady ionic current as a detection mechanism. As shown in Figure 67, air inside a sensing chamber is ionized by a 1 μCi [241]Am source. Whenever smoke particles enter the sensing chamber, the ionic current is reduced. This drop in current is sensed by a specialized circuit that triggers an alarm. Many semiconductor companies offer very inexpensive, one-chip solutions for ionization smoke detectors. Examples include the Freescale MC14468, Microchip RE46C114, and Allegro A5367 integrated circuits.

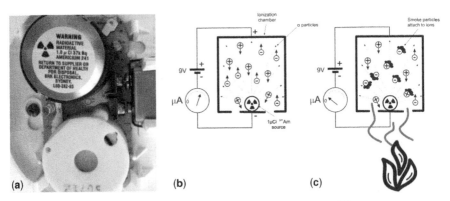

Figure 67   Ionization-type smoke detectors contain a 1 μCi source of [241]Am. **(a)** The americium source is a small button placed inside the ionization chamber. **(b)** Alpha particles from the [241]Am ionize the air within the chamber, allowing a current to flow between its two electrodes. **(c)** When smoke particles enter the chamber, they cling to the ions, causing a marked drop in current flow, which triggers the alarm. Please note that it is not legal in the United States (and some other countries) to remove the radioactive source from a smoke detector.

WHAT ARE ALPHA PARTICLES?

In 1909, Rutherford and his colleague Thomas Royds really wanted to know what made up alpha particles. They carried out many experiments with alpha particles emitted from radon—a gas that occurs naturally as the decay product of radium, and which often accumulates in basements to a point where it may pose health risks. The radon samples often contained helium, making Rutherford and Royds suspect that alpha particles were "naked" helium ions, that is, helium atoms that had lost two electrons. They obtained definitive proof after conducting a very elegant experiment (Figure 68) in which they placed a large quantity of purified radon gas into a glass tube with an extremely thin wall. An evacuated collecting tube surrounded the thin-walled, radon-containing tube. After 2 days, they passed high voltage between two electrodes to cause any collected gas to glow—just like what happens inside a

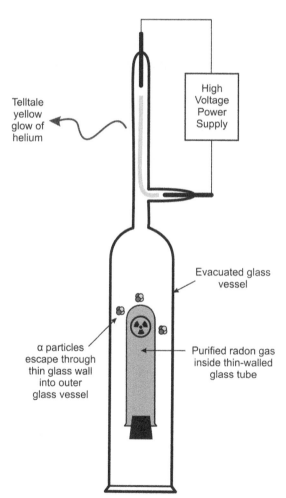

Telltale
yellow
glow of
helium

High
Voltage
Power
Supply

Evacuated glass
vessel

α particles
escape through
thin glass wall
into outer
glass vessel

Purified radon gas
inside thin-walled
glass tube

Figure 68   This is a highly simplified view of the experiment performed by Rutherford and Royds to demonstrate that alpha particles are helium ions. They placed radon gas—an alpha emitter—inside a glass tube with very thin walls. This tube was itself inside an evacuated collecting tube. After 2 days, they could detect helium gas in the collection tube by watching for the telltale yellow glow that helium produces when excited by high voltage.

modern neon tube. Indeed, the contents produced a telltale yellow glow indicative of pure helium.

As a control, Rutherford and Royds also ran the experiment with pure helium gas inside the thin-walled tube. They left it for a few days and attempted again to detect helium gas that would have diffused through the thin glass wall into the outer collection vessel. However, they could not detect any helium, meaning that for some reason an alpha particle can penetrate the thin glass wall of the tube, but a helium atom cannot. Rutherford thus concluded that the "naked" helium ions were much, much smaller than regular helium atoms. This hinted that the atom was not as homogeneous as the "plum-pudding" model would have it, and that the mass might be more concentrated, leaving quite a bit of empty space within the atom's boundaries.

RUTHERFORD'S ALPHA-SCATTERING EXPERIMENT

Meanwhile, Rutherford's colleague Hans Geiger was exploring the use of thin films of mica—a transparent mineral that can be split into very thin slices—as radiation windows for his detector tubes. Geiger noticed that mica spread out a beam of alpha radiation by slightly deflecting some of its alpha particles. Together with Ernest Marsden—a young undergraduate student working for Rutherford—Geiger found that weak reflections could be detected from a beam of alpha particles directed toward some metals. Although the "plum-pudding" model of the atom could explain some of the alpha-particle scattering in mica, one wouldn't expect to see any alpha particles to bounce back when hitting the gelatinous "pudding."

Armed with these clues, Rutherford designed an experiment to measure the scattering and reflections of alpha particles using an extremely thin gold leaf. Figure 69a shows a simplified diagram of Rutherford's apparatus. In it, a stream of alpha particles was aimed at an extremely thin (under 0.1 µm) gold foil. Geiger and Marsden spent

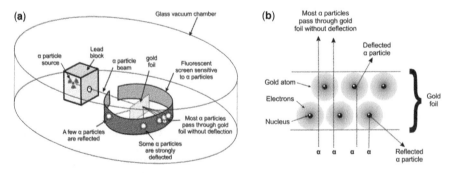

Figure 69   In Rutherford's scattering experiment, (a) when a beam of alpha particles is directed at a thin gold foil, most particles pass through the foil undeflected, but some are deflected at an angle, and a few even reflect back toward the particle source. (b) This could not happen unless atoms are made mostly of empty space, and only the alpha particles that graze or strike a small, solid nucleus are deflected.

months in the darkroom counting scintillations on a fluorescent screen that could be rotated around the foil. They found that almost all of the alpha particles went straight through the gold foil, continuing on a straight-line path until they hit the fluorescent screen. Some of the alpha particles were deflected slightly, usually $2°$ or less. However, just a very few alpha particles—around one in 20,000—bounced off the foil at an angle of $90°$ or more.

Rutherford found the latter to be completely incompatible with the "plum pudding" model. As he is often quoted: "It is as if you fired a 15-inch artillery shell at a sheet of tissue paper and it came back to hit you!"

Rutherford's solution to the riddle of large- and small-angle scattering was to concentrate the atom's mass in a very small space we now call the *nucleus*. While Thomson's "plum-pudding" model spread the entire mass of the atom throughout a sphere of radius 10^{-10} m (Figure 55), Rutherford placed most of the mass at the center of the atom, in a space much smaller than the gold atom itself, which he calculated to be smaller than 34×10^{-15} m in radius. We now know that the radius of the gold atom is about 7.5 fm (femto $= 10^{-15}$), while the radius of a single proton (the nucleus of a hydrogen atom) is only 0.88 fm.

As shown in Figure 69b, the nucleus is so small that in the highest probability an alpha particle would just fly through the gold foil as if nothing were there. Some alphas passed near some gold atom nuclei, and were slightly deflected. The gold sheet used in the experiment was some 50 atoms thick, so some of the alpha particles were deflected by more than one close encounter with a gold nucleus, causing scattering of up to $2°$. However, a very, very few alpha particles hit a nucleus smack-on. Since the alpha particle is positively charged, it is repelled by the equally charged nucleus with such force that it may bounce back toward its source.

Replicating Rutherford's experiment in its original form requires an enormous amount of patience and dedication. Geiger and Marsden's months in the darkroom counting scintillations through a microscope eyepiece was so traumatic, it is said, that Geiger's motivation for developing the GM tube was to replace his tired eyes by an automatic detector.

The following is an experiment that you should definitely try so that you can empathize with Geiger's aggravation: purchase a small screen coated with silver-activated zinc sulfide and look at it through a magnifying glass or jeweler's loupe. After adapting your eyes to darkness, you will be able to see the faint scintillations produced on the screen when it is exposed to a source of alpha particles. Each flash of light is the result of a single alpha particle interacting with the activated ZnS. Back in 1903, Sir William Crookes (remember him?) built the first device of this kind and named it *spinthariscope.*

You can buy the screen for less than \$15 from GeoElectronics, and use one of the sources of alpha radiation that we have discussed. For example, you could use a ^{210}Po disk source, a lantern mantle containing ^{232}Th, or the small button containing ^{241}Am from a smoke detector (but only if you are in a country that allows you to take one apart legally). You could also purchase a ready-made spinthariscope, complete with a tiny radioactive speck, from United Nuclear.

If catching the flashes produced by a source just under the zinc sulfide screen requires effort, just imagine what it takes to catch one in 20,000 particles in the direction of the backscattered radiation! We are sure that you now have a whole new appreciation for the devotion Geiger and Marsden had for their work.

There are a few modern ways of conducting Rutherford's scattering experiment in a more convenient way. One popular device used in many colleges is based on a design by Charles W. Leming of Henderson State University in Arkansas. It uses a special plastic-sensitive film to record particle hits. The film is exposed inside a vacuum chamber for at least 7 days, after which it is "developed" in a sodium hydroxide (caustic soda) solution. A stereo microscope is then used to look at the film and count alpha particles deflected by up to $\pm 20°$.

The film is made by Kodak-Pathé in France (type LR115-II),[22] and consists of a 13-μm-thick red layer of cellulose nitrate over 100-μm-thick clear polyester film. Alpha particles damage the nitrocellulose, and the film can be etched to reveal the impact spots by soaking it for 24 hours in a 2.5 molar solution of NaOH[III] at a temperature of 40°C. The individual pits can be viewed under a microscope at $50–200 \times$ magnification. The film is rather expensive, but can be purchased from the Science Source (catalog number EN-21) as replacement film for their commercial version of Leming's apparatus for studying alpha scattering.

The now-extinct "Amateur Scientist" column in the February 1986 issue of *Scientific American* described the amateur device built by Rudy Timmerman of Wickes, Arkansas, to measure alpha-particle scattering using this method.[59] Figure 70a shows the basic configuration used in Timmerman's design, as well as in the commercial Rutherford Scattering Apparatus made by Daedalon (Science Source model EN-20). Typical results are shown in Figure 70b. Counts are normalized to those at 2.5°, since the number of hits within this scattering angle is simply enormous. The curve agrees with the data originally obtained by Geiger and Marsden, as well as with the formula that Rutherford worked out for the scattering of alpha particles by gold nuclei.[20]

Our homemade version of the apparatus is shown in Figure 71. We use two radioactive sources to produce two fine alpha-particle beams. However, only one of the two alpha-particle beams undergoes Rutherford scattering, allowing direct comparison between alpha-particle dispersion due to beam divergence and due to Rutherford scattering.

The availability of alpha-particle sources that can be purchased without special licensing requirements is very limited. One popular source used by experimenters is a 1 μCi ^{241}Am button extracted from an ionization-type smoke detector. However, this practice is illegal in the United States. For this reason, the Daedalon Rutherford scattering apparatus uses an exempt ^{210}Po source with a maximum activity of 0.1 μCi. As an alternative, we use a static eliminator cartridge (StaticMaster) that contains two metallic foils plated with ^{210}Po. We make legal use of the sources within the cartridge (with a total activity of 500 μCi), since we do not modify the cartridge in any way. It must be noted that the predominant 5.30 MeV emission from ^{210}Po has a half-life

---

[III]WARNING! Strong solutions of sodium hydroxide will cause chemical burns, permanent injury, or scarring if they contact unprotected human tissue. They will cause blindness if they contact the eye. Protective equipment such as rubber gloves, safety clothing, and eye protection should always be used when handling these solutions.

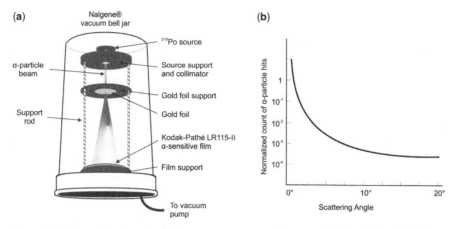

(a)

Nalgene® vacuum bell jar

²¹⁰Po source

α-particle beam

Source support and collimator

Gold foil support

Support rod

Gold foil

Kodak-Pathé LR115-II α-sensitive film

Film support

To vacuum pump

(b)

Normalized count of α-particle hits

10^1

10^{-1}

10^{-2}

10^{-3}

10^{-4}

0°   10°   20°

Scattering Angle

Figure 70   A popular educational apparatus to measure alpha scattering by a thin gold foil uses this simple setup. **(a)** Alpha particles from a ²¹⁰Po source at the top of the chamber are collimated into a fine beam by an aluminum disk with a small hole. The beam hits a thin foil of gold leaf. An alpha particle–sensitive film records impacts by direct and scattered particles. The stack is placed inside a plastic bell jar, which is kept at a moderate vacuum for 7 days. The figure is not to scale in order to show detail. **(b)** A graph of particle strikes that can be counted under a microscope. These strikes appear on the film as clear spots in a red background after development in a caustic soda bath.

of 138.3 days, so the cartridge would need semi-annual replacement if the setup is used as part of a regular didactic lab course.

     We purchased 23-karat gold leaf at a gourmet cooking store, where it is sold as thin leaves for use as decorations on foods. The ones we used were made by Fabbriche

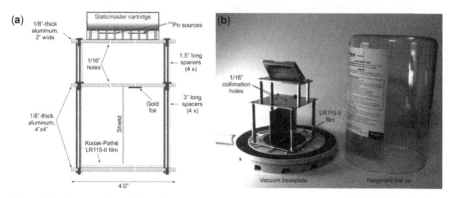

(a)

1/8"-thick aluminum, 2" wide

Staticmaster cartridge

²¹⁰Po sources

1/16" holes

1.5" long spacers (4 x)

Gold foil

3" long spacers (4 x)

1/8"-thick aluminum, 4"x4"

Shield

Kodak-Pathé LR115-II film

4.0"

(b)

1/16" collimation holes

LR115-II film

Vacuum baseplate

Nalgene® bell jar

Figure 71   Our version of the alpha-scattering apparatus uses an unmodified Staticmaster® cartridge to produce two fine alpha-particle beams. **(a)** Only one of the two alpha-particle beams undergoes Rutherford scattering, allowing direct comparison between alpha-particle dispersion due to beam divergence and due to Rutherford scattering. **(b)** The apparatus must be operated inside a vacuum chamber during exposure, as alpha particles have very limited range in air, but the vacuum requirements are not stringent. A student-grade polycarbonate vacuum chamber is sufficient.

Riunite Metalli in Foglie e in Polvere S.p.A. (Morimondo, Italy). It must be noted that handling the gold leaf and attaching it to the holder plate requires extreme care. The gold leaf is so thin that it shreds at the lightest touch. It is best handled with a brush that is charged with some static electricity by rubbing it lightly on the skin. We placed a few tiny drops of model-airplane glue on the frame, and allowed these to grab hold of the gold leaf. We use two layers of this leaf to make it thick enough to yield easily observable alpha-particle scattering.

We fabricated the frame for the apparatus out of $\frac{1}{8}$-in.-thick aluminum sheet. We drilled the collimation holes with a $\frac{1}{16}$-in. bit. We also drilled coaxial holes on the film support plate through which we could shine a laser pointer to aid in aligning the aluminum plates. A thin aluminum sheet shields the part of the alpha particle–sensitive film recording the unscattered beam from alpha particles that may be scattered by the gold foil at wide angles.

The apparatus must be operated inside a vacuum chamber during exposure, because alpha particles have very limited range in air. The vacuum requirements for this setup are not stringent at all. The range of the ^{210}Po 5.30-MeV alpha particles at atmospheric pressure is approximately 3.8 cm. Since the kinetic energy loss experienced by alpha particles scales linearly with pressure, evacuating the chamber to approximately $\frac{1}{1000}$ of normal atmospheric pressure ensures that alpha particles can travel virtually unimpeded by air between the source and the LR115-II film. In addition, since this film is not sensitive to light or other types of radiation besides alpha particles and neutrons (such as beta or gamma rays), the chamber can be transparent without affecting the results of the experiment. As such, a student-grade polycarbonate vacuum chamber (e.g., 4.7-L Thermo Scientific Nalgene$^{®}$ vacuum chamber) suffices for this experiment.

We connected the vacuum chamber through a short length of vacuum-service rubber tubing to a hose-to-KF16 vacuum flange adapter. We terminated the KF16 port with a vacuum valve so we could periodically lower the pressure in the vacuum chamber using our refrigeration-service vacuum pump (Robinair model 15600). We also took pressure readings twice every day to make sure that the chamber was holding its vacuum below 1 Torr.

A portion of our etched LR115-II film is shown in Figure 72. The spot on the left-hand side of the picture was produced by the collimated beam, while the one on the right was produced when the alpha particles underwent Rutherford scattering. The number of scattered particles drops rapidly with increasing scattering angle, so we

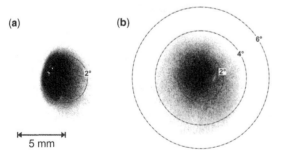

(a) **(b)**

5 mm

Figure 72   Alpha-particle hits on the Kodak-Pathé LR115-II film are shown after it is etched for 24 hours in 2.5 N NaOH at 40°C. **(a)** The collimated beam. **(b)** Alpha-particle beam that has been scattered by a thin gold film.

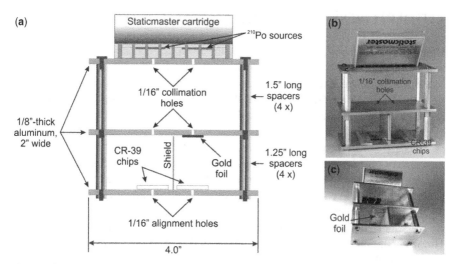

Figure 73   Inexpensive CR-39 plastic dosimetry chips can be used instead of the Kodak-Pathé LR115-II film to record the collimated and Rutherford-scattered alpha particles. (a) The dimensions given here allow alpha particles to be recorded out to a scattering angle of $\theta = \pm 17°$. (b) The setup is sufficiently small to fit within a small student-grade polycarbonate vacuum chamber. (c) Two layers of 23-karat gold leaf are tacked with tiny drops of model-airplane glue.

show only scattering out to 6°. However, the setup is able to register scattered particles out to 20°.

A lower-cost alternative to LR115-II film is plastic chips that are commonly used in neutron dosimeters and home radon test kits. The chips are of a plastic known as CR-39, which is commonly used in the manufacture of eyeglass lenses. When exposed to neutrons or alpha particles, the plastic is microscopically damaged at the impact sites. Just as with the LR115, the pits can then be enlarged sufficiently to be viewed with a $200\times$ microscope by etching the chips in a caustic solution of sodium hydroxide (6 N NaOH at 70°C for 7 hours). We purchased the CR-39 dosimeter chips from Landauer for $3 apiece. Each chip measures 19 mm $\times$ 9.4 mm $\times$ 0.9 mm and is individually etched with a serial number for identification. As shown in Figure 73, one of the chips records alpha particles that spread only due to beam divergence after collimation. The other chip records alpha particles scattered by the gold foil within the range of $\theta = \pm 17°$. Figure 74 shows our results after a 7-day exposure.

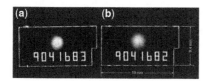

Figure 74   Alpha-particle hits on CR-39 plastic chips after etching for 7 hours in 6 N NaOH at 70°C. (a) The collimated beam. (b) Alpha-particle beam that has been scattered by a thin gold film.

RUTHERFORD'S PLANETARY MODEL OF THE ATOM

Rutherford came up with a new model of the atom that accounted for his scattering results: he proposed that the positive charge and its associated mass is concentrated in the nucleus, and that the negatively charged electrons are orbiting around it, like planets. In his 1911 paper discussing the alpha-particle scattering results,[20] Rutherford mentioned the planetary atomic model that had been proposed back in 1904 by Japanese physicist Hantaro Nagaoka.

Nagaoka was inspired by the structure of Saturn, and imagined the atom could have a very massive nucleus, analogous to the massive nature of Saturn (Figure 75). Electrons would revolve around the nucleus bound by electrostatic forces, just like Saturn's rings revolve around the planet bound by gravitational forces.

Nagaoka believed that such a system would be stable, analogous to Saturn's rings, which remain steady because they orbit a very, very massive planet. However, the problem with this assumption is that electrons, unlike Saturn's rings, are electrically charged.

The issue is that the electrons in the planetary model move in circles. You may remember from basic physics that if a body is moving around a circle, even if it is moving at a constant speed it is accelerating. This is because it is changing direction (it isn't moving in a straight line). Now, Maxwell's equations of electromagnetism predict that accelerating electric charges should emit electromagnetic waves. As such, the electrons in the planetary model should lose energy by radiating electromagnetic waves. Then, like a satellite losing energy because of friction with Earth's upper atmosphere, the electrons should quickly spiral to their death into the nucleus (Figure 76). All atoms in the universe would have disappeared just an instant after their creation!

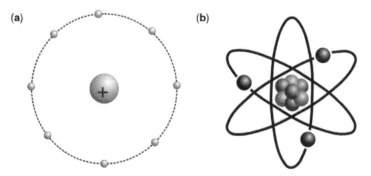

Figure 75   The planetary model of the atom. (**a**) In 1904, Hantaro Nagaoka proposed a model of the atom based on an analogy with the rings of Saturn. He proposed the existence of a very massive nucleus (analogous to a very massive planet), and electrons revolving around the nucleus, bound by electrostatic forces (analogous to the rings revolving around Saturn, bound by gravitational forces). This model was mentioned by Rutherford in his 1911 paper, in which he announced the discovery of the nucleus. (**b**) Despite its inaccuracy, the Rutherford model caught the imagination of the public, and is still used as a symbol for atoms and atomic energy.

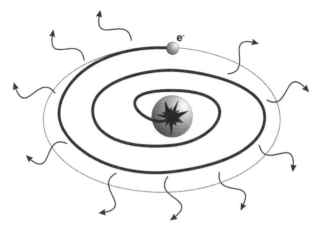

Figure 76 The planetary model of the atom is inherently unstable. This is because an orbiting electron is an electric charge under acceleration, and must therefore radiate electromagnetic energy at the cost of centrifugal force. The electron in this model would thus spiral to its death in a very small fraction of a second.

In the next chapter, we will see how quantum physics comes to the rescue by explaining how electrons may nevertheless remain in stable orbits around the nucleus.

EXPERIMENTS AND QUESTIONS

1. Build the glow-discharge tube of Figure 42 and connect it to your vacuum manifold (which should include a T/C gauge). Connect the negative high-voltage terminal of the power supply of Figure 39 to the $\frac{3}{8}$-in. aluminum rod, and the power supply's ground terminal to the vacuum port. Dim the room lights, turn on the high voltage (to a voltage of around 10–15 kV), and start evacuating the system. Record what happens at different pressures. At what pressure does the glass next to the cathode start to produce a greenish glow? At what pressure does this greenish glow appear close to the anode? What causes this glow?

2. Build the "Maltese Cross" CRT of Figure 43. Use the electron gun shown in Figure 41. Connect the tube to your vacuum pump. Connect the negative high-voltage terminal of the power supply of Figure 39 to the cathode rod, and the power supply's ground terminal to the hollow anode through a thin wire leading to the vacuum port. You should also connect the "happy face" electrode to the ground terminal. Dim the room lights, start evacuating the system, and turn on the high voltage. At some point you should see the shadow of your "happy face" electrode on the phosphor screen. Is the shadow sharp or fuzzy? What is the significance of this observation?

3. Bring the north face of a strong magnet close to the CRT's screen flask. Observe what happens and then turn the magnet around so that the south face is close to the screen. What effect does the magnet have on the "happy face" shadow? What is the effect of changing the magnet's polarity? Why?

4. Reverse the polarity of the power supply so the "cathode rod" of the "Maltese Cross" tube is connected to the power supply's positive high-voltage terminal; and the vacuum system's port, along with the "happy face" electrode, are grounded (and negative in relationship to the "cathode rod"). Does the screen glow so that a shadow can be seen? Why?

5. Disconnect the "happy face" electrode from ground and energize the tube. Does the shadow become distorted? Why?

6. Retract the "happy face" electrode on the side port of the flask so it lies flat against the flask wall, as shown in Figure 45b. Connect a digital multimeter with 10 MΩ input impedance between the "happy face" electrode's post and ground. Set the multimeter to measure volts DC, which will make it measure the current through the "happy face" electrode, where the measurement in microamperes is equal to the voltmeter's reading divided by 10. Evacuate the tube and turn on the power supply until a clearly visible spot shows on the screen. Stabilize the pressure and voltage to maintain that spot. Measure the voltage on the digital multimeter. Bring a magnet close to the tube so that the cathode rays are bent toward the "happy face" electrode. What happens to the multimeter reading when the cathode rays reach the "happy face" electrode? Why? What is the polarity of the charge conveyed by the cathode rays? What is the amount of charge arriving at the collection electrode (use $I[A] = Q[\text{Coulomb}]/s$)? What were Thomson's conclusions when he conducted his version of Perrin's experiment?

7. Build the CRT shown in Figure 47. Connect one of the plates in the vertical pair to ground (the vacuum pump manifold). In addition to the high-voltage power supply driving the electron gun, connect a second high-voltage power supply (e.g., PMT bias power supply) between ground and the other deflection plate. Vary the deflection voltage as the CRT is evacuated and record at what pressure it is possible to deflect the beam electrostatically. Once you locate the region where deflection is possible, stabilize the pressure and electron-gun voltage to produce a steady spot on the screen. Observe the deflection of the beam as a function of deflection voltage. Invert the polarity of the deflection power supply and run the experiment again. What is the relationship between cathode-ray beam deflection and the potential difference between the plates? What did Thomson conclude when he conducted his experiment? Why wasn't Hertz able to observe electrostatic deflection when he performed a similar experiment?

8. Build the simple oscilloscope of Figure 49. Connect two isolated digital voltmeters to measure: (1) the voltage between the CRT's cathode and anode electrodes, and (2) the voltage between the two vertical deflection plates. Measure the deflection of the spot on the screen as a function of voltage between the deflection-plate electrodes. What is the relationship between them? What is the velocity and kinetic energy of the electrons (based on the accelerating potential between the cathode and anode)? What can you say about the force on the electrons produced by the deflection field?

9. Build the setup to measure e/m using Hoag's method, as shown in Figure 52. Set the accelerating voltage to an initial value $V = 300$ V, and the deflection controls to produce a thin vertical line. Increase the current through the solenoid until a full turn is traced by the line on the CRT. Record the accelerating potential and current through the solenoid. Repeat at various accelerating potentials in steps of 100 V up to a maximum of 800 V. Calculate $e/m = 8\pi^2 V/B^2 L^2$ for each run and yield an average estimate for e/m. How well do your measurements compare with the accepted value of $e/m = 1.76 \times 10^{11}$ Coulomb/kg? What are possible sources of error? Explain.

10. What is the velocity and kinetic energy of the electrons at the different accelerating potentials in this experiment?

11. In what way does the magnetic field used in Hoag's method behave like an optical lens? What is the effect of the strength of the magnetic field and the electron's velocity on the focusing power of this "electron lens"?

12. Build the experimental setup to measure e/m using a "magic-eye" tuning tube as shown in Figure 54. Power the tube with 130 V, and record the solenoid currents required to bend the V-shaped shadow to match the bend radius of cylindrical $\frac{1}{4}$-in., $\frac{1}{2}$-in., and $\frac{3}{4}$-in. wooden dowels. Repeat the measurements at tube voltages between 90 V and 190 V. Estimate e/m based on your results. What are possible sources of error? Why can only the ratio of e/m, but not the exact values of e or m be measured using this setup?

13. Assuming $e = 1.60 \times 10^{-19}$ Coulomb, use $m_e = B^2 eR^2/2V$ to estimate the mass of the electron. How well does your estimate match the accepted value of $m_e = 9.11 \times 10^{-31}$ kg?

14. What did Thomson conclude from his experiments about the negatively charged particle of cathode rays? How did he reconcile his discovery with the fact that atoms are usually neutral?

15. Set up a radiation counter with a GM tube that is sensitive to alpha, beta, and gamma radiation. Use exempt, plastic-disk sources containing Polonium-210 (^{210}Po), Strontium-90 (^{90}Sr), and Cobalt-60 (^{60}Co) to experiment with the penetrating power of alpha, beta, and gamma radiation in air. Measure how radiation counts decay for each of the sources as the distance between the source and the GM tube window is increased between 1 cm and 1m. Select the sampling distances for each source, remembering that alpha is very short range, beta has a range of about 10 cm in air, and gamma gets weaker with distance but doesn't come to a stop at any particular distance. Assuming that the sources emit equally in all directions, calculate the effect of air's density (~ 1.3 mg/cm^3) in attenuating each of the radiation types. Remember to correct for background radiation counts.

16. How do the counts detected from a source vary as a function of distance?

17. Try to find the narrow range of distances at which the counts produced by your alpha-particle source abruptly drop to zero. The "mean range" is midway between the point at which counts first drop off sharply and the first point at which counts above background are zero. Why don't alpha-particle counts decay smoothly to zero as distance is increased?

18. Place the ^{210}Po, ^{90}Sr, and ^{60}Co sources—one at a time—at an appropriate distance from the GM tube window to give you a clear count (e.g., 1–3 cm, depending on the source). Use absorbers of various materials and thicknesses (e.g., paper, aluminum foil in the range of 0.01–0.03 mm, aluminum sheet in the range of 0.5–10 mm, and lead sheet in the range of 1–10 mm) to determine their effectiveness as alpha-, beta-, and gamma-radiation shields. Correct your data for background radiation counts. Graph the counts detected from each source when different shielding materials and thicknesses are used. Calculate the absorber thickness for each shield (the thickness for an absorber is given in mg/cm^2, that way its density is taken into account), and graph the attenuation of each radiation type as a function of absorber thickness. What general statement could you make about the thickness of the absorbing material on the count rates?

19. Set up an exempt, plastic-disk ^{90}Sr beta-particle source, two copper washers, and a GM tube, as shown in Figure 65. Bring a strong magnet close to the setup, as shown in the same figure. Is there a point at which the beta-particle beam is bent by 90° so that it can be detected by the GM counter at a right angle? What magnetic polarity is required to bend the beam in what direction? What does this tell us about beta particles? Conduct the same experiment with a ^{22}Na β^+ positron particle source. Did you obtain similar results? Explain.

20. Build the experimental apparatus of Figure 71 (or, alternatively, the one of Figure 73) to measure alpha-particle Rutherford scattering. Expose the alpha particle–sensitive film for 7 days under a vacuum below 1 Torr, and then etch the film as described in the text. Use a microscope or projector to magnify the film. What is the difference between the spot produced by the collimated beam and the one produced when the alpha particles undergo Rutherford scattering? Count the number of hits within 2° intervals for angle θ out to 20°. Taking into consideration the results that you obtain for the collimated beam, how do your results relate to Rutherford's prediction that alpha-particle scattering depends on $\csc^4(\theta/2)$?

21. What atomic model did Rutherford adopt to explain his alpha-particle scattering experiments? What is the problem with this model?

CHAPTER 4

THE PRINCIPLE OF QUANTUM PHYSICS

The last chapter left us with a planetary model of the atom that takes into account the existence of the nucleus, but which is inherently unstable. To recap, the model proposes that the centrifugal force of the revolving electron just exactly balances the attractive force of the nucleus. However, the issue is that the electrons—being electrically charged—must radiate energy as they move in a circular orbit. This is because Maxwell's equations of electromagnetism predict that accelerating electric charges must emit electromagnetic waves. If the electrons radiate energy, they would lose velocity, thus spiraling into the nucleus.

Even if an electron's orbit could somehow be stabilized, a related puzzle became a real quandary: it had been known since the late nineteenth century that excited atoms of a single element do not radiate a continuous spectrum. Instead, they produce a discontinuous spectrum of many lines. That is, as shown in Figure 77, instead of a smooth spectrum containing all colors, the light emitted by excited atoms consists of some number of discrete waves of different wavelengths. Each element has an individual, characteristic line spectrum, called its *emission spectrum.*

In 1862, Anders Ångström discovered three lines in the spectrum of hydrogen, and later found a fourth line. By 1871, he had measured all four wavelengths to a high degree of accuracy: one red line (6,562.852 Å = 656.2852 nm), one blue-green line (4,861.33 Å = 486.133 nm), and two violet lines (4,340.47 Å = 434.047 nm and 4,101.74 Å = 410.174 nm).

H_α, H_β, H_γ, and H_δ are the official designations for the four hydrogen lines of the visible portion of the spectrum. However, there are others in other parts of the electromagnetic spectrum. The group of visible hydrogen spectral lines is called the Balmer Series in honor of the Swiss mathematical physicist Johann Jakob Balmer, who was able to calculate the four visible lines' wavelengths using one formula, now called the Balmer Formula.

EMISSION SPECTROSCOPY

Let's see those hydrogen lines for ourselves!

We will use the kind of hydrogen "spectrum tube" available at any store that sells educational scientific supplies. Although they are all pretty much the same,

Exploring Quantum Physics Through Hands-On Projects. By David Prutchi and Shanni R. Prutchi
© 2012 John Wiley & Sons, Inc. Published 2012 by John Wiley & Sons, Inc.

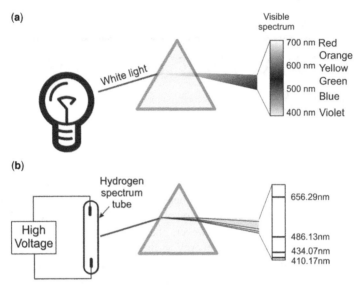

Figure 77 (a) The spectrum of light produced by white light is continuous, containing all colors of the visible spectrum. (b) The light produced by excited hydrogen is discontinuous, having just a few, well-defined spectral lines.

their price varies quite a bit based on the store's brand name. The least-expensive spectrum tubes are sold by Information Unlimited (under $20). You can buy these filled with different gases, such as hydrogen, oxygen, nitrogen, neon, argon, and helium. They are made specifically to study the emission spectra of pure elements, so their internal pressure, electrode materials, and dimensions have been optimized to produce sufficient intensity for spectral analysis. These tubes are not designed to be used as decorative lamps, and should not be operated for longer than 30 seconds before they are switched off for at least another 30 seconds. You can energize these tubes from the power supply shown in Figure 78. The high-voltage module we used is sold by Information Unlimited as model NEON21, but the high-voltage AC power supply of Figure 39a (without the voltage multiplier of Figure 39b) works just as well. Please note that it is usually enough to connect only the high-voltage AC terminal from a high-frequency power supply to one of the tube ends, leaving the other tube end floating. If this works with your power supply, it will prolong the tube's life.

After you experiment with the purpose-made tubes, try building your own spectrum tubes using the simple glow-discharge tube (Figure 42) by back-filling the tube with some hydrogen, helium, or other gases to which you may have easy access. For example, you could produce hydrogen by electrolysis of water, use helium from a party balloon, or obtain argon from a welding supplies store. However, don't expect the spectral purity of commercial spectrum tubes unless you have a system that can pull a very deep vacuum and use laboratory-grade gases. This is because your system at a rough vacuum will still contain quite a bit of air (which is mostly nitrogen), water vapor, and by-products of the vacuum oil.

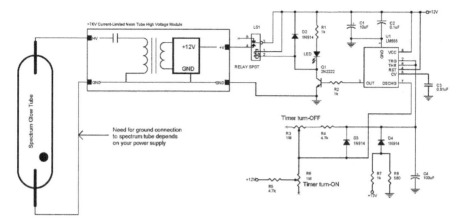

Figure 78   We built this simple high-voltage power supply to energize commercial spectrum tubes. R3 and R6 should be trimmed to power the tube for only 30 s before it is turned off and allowed to cool down for 30 s. We built a plastic stand for the tube and used clips from a $\frac{1}{2}$-in.-diameter fuse holder to connect the power supply to the tube.

The other item that we need to observe hydrogen's emission lines is a *spectrometer* or *spectroscope*.* We built one using a prism back in chapter 2 to study the blackbody radiation spectrum (Figure 21). However, one that is more portable and easy to use can be made from a grooved piece of plastic film known as a diffraction grating. The fine periodic grooves in the grating diffract light at an angle that is dependent on wavelength. For example, take any compact disk or DVD, and shine a light on its surface. The very fine grooves arranged in a spiral act as a grating and disperse the light into its constituent components, because each wavelength that is part of the illuminating white light is sent into a different direction, producing a rainbow of colors under white light illumination. The end result is similar to the operation of a prism, although the optical mechanism is very different.

There are many plans on the Web for building a spectroscope with a CD acting as a reflection diffraction grating. However, we recommend that you use a transmission diffraction grating slide for making your spectroscope. These are available from educational science supply stores and many other online sources, often for under $1. Chose a linear diffraction grating with 1,000 lines/mm that is mounted on a 35-mm slide. Alternatively, you may purchase a ready-made, inexpensive student spectroscope with a calibrated wavelength scale, preferably the Project STAR spectrometer manufactured by Science First.

To make your own spectroscope, find or make a cardboard box for the body of your instrument. An empty cereal box works very well. Carefully take the box apart and paint the inside flat black. Once it dries completely, cut a rectangular hole for the diffraction grating slide, as shown in Figure 79. Cut a very narrow slot directly

---

*The distinction between the two is not clear-cut. For purists however, spectroscopy is the overall term for the study of the interaction between radiation and matter as a function of wavelength, while spectrometry relates to the quantitative measurement of spectroscopic observations.

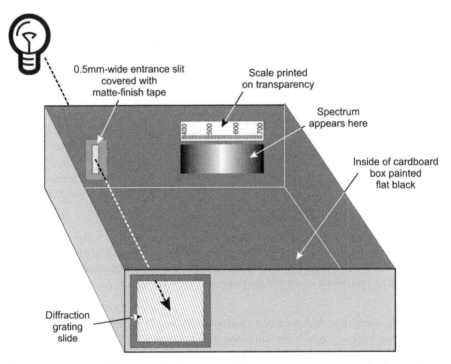

0.5mm-wide entrance slit
covered with
matte-finish tape

Scale printed
on transparency

Spectrum
appears here

Inside of cardboard
box painted
flat black

Diffraction
grating
slide

Figure 79   You can build a simple, but very useful, spectroscope using a diffraction grating slide. The spectroscope is shown with the top cover removed for clarity. Make the housing out of a cardboard box the size of a cereal box. Cut a narrow entrance slit to produce a very sharp spectrum. Print a calibrated scale on a piece of translucent plastic above the spectrum so that you can explore sources with unknown spectra.

across the slide. The slot should be around 0.5-mm wide and 3-cm tall. If it's easier for you, cut a wider slot (e.g., 1 cm) and use two index cards over the slot to form the narrow slit. Place a single strip of translucent adhesive tape (e.g., Scotch Matte Finish Magic Tape) over the slit.

Next, temporarily reassemble the box with the exception of the top cover. Tape the diffraction grating slide in place, and look at a light source. Mark the area where you see the spectrum, and cut a 1-cm high slot over that area. Tape a piece of translucent plastic (we used a strip of plastic that we cut out of a milk jug) over the slot that you just cut. This is where you will print your wavelength scale. Now, observe a number of sources with known emission spectra (e.g., the LEDs you used in chapter 2) and mark the scale accordingly. Lastly, tape the top of the box and start exploring!

Observe various sources of light that are used for everyday illumination. What do you notice about the spectrum produced by an incandescent lightbulb, a fluorescent lightbulb, and a street lamp? By now you should be able to tell why the incandescent lightbulb behaves as a blackbody and produces a continuous spectrum. But how about the fluorescent light? Why does it look white to you despite its spectrum consisting of discrete lines?

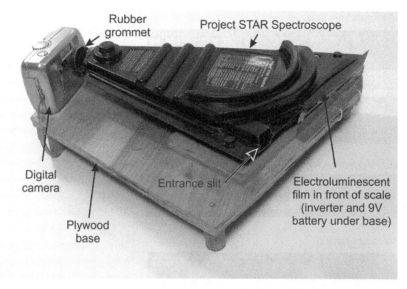

Figure 80   We coupled an inexpensive digital camera and a piece of EL foil to a Project STAR spectrometer to make it possible to capture light spectra for computer analysis.

Adding an inexpensive digital camera to the spectroscope allows the "rainbow" spectra to be analyzed by a computer to produce quantitative spectral plots. As shown in Figure 80, we built a wooden base to hold our Project STAR spectrometer and an old digital camera. We used a rubber grommet to couple the spectrometer's eyepiece to the camera's lens. We taped a piece of electroluminescent (EL) foil in front of the scale slot so that we would always get our pictures to include an evenly illuminated scale. We mounted the inverter for the EL lamp and a 9-V battery under the plywood base. We used some EL foil that can be cut to the desired shape with scissors. These EL foils are made by Miller Engineering, and are sold in hobby stores to illuminate model railroad billboards. However, you could also pull out the LCD display EL (or LED) backlighting from an LCD clock or calculator. Figure 81 shows a few examples of spectral images obtained with our simple setup.

We convert pictures taken with this spectrometer into spectral plots with VisualSpec, a free program created by Valérie Desnoux.[†] This program requires that spectral images be loaded in either .pic or .fit (the "Flexible Image Transport" standard that is commonly used by NASA to store and transfer astronomical images) format, so we convert the .jpg images produced by our camera using one of the free online image conversion tools. VisualSpec is very well documented, and includes excellent user instructions.

Now, turn on your hydrogen spectrum tube and observe the light through your spectroscope. Can you distinguish the Balmer lines? Well, in 1885, Balmer noticed that the lines get closer together, following a specific pattern as wavelength

[†]The VisualSpec software by Valérie Desnoux is available for free download at http://astrosurf.com/vdesnoux.

Helium spectrum tube

Neon spectrum tube

Mercury spectrum tube

Fluorescent lighting

Solar spectrum

Figure 81 Spectral images obtained with our simple spectrometer of Figure 80. Top scale is in electron-volts [eV]; lower scale is in nm.

increases. Balmer figured out that lines appeared with regularity at certain multiples of 364.56 nm. He found the empirical formula:

$$\lambda = B\left(\frac{m^2}{m^2 - 4}\right)$$

where λ is the wavelength of a hydrogen line, B is Balmer's constant with the value of 364.56 nm, and m is an integer greater than 2. Using this formula, Balmer was able predict lines in the invisible portion of the spectrum, which were later confirmed when instruments were invented to measure lines outside the visible spectrum.

In the 1880s, the Swedish physicist Johannes Rydberg was also working on the spectra emitted by the elements. He found out that he could simplify his calculations by using the wavenumber (the number of waves occupying a set unit of length, equal to $1/\lambda$, the inverse of the wavelength) as his unit of measurement. When Balmer's paper was published, Rydberg generalized the Balmer equation to include new hydrogen lines that were unknown when Balmer did his work (Figure 82). Rydberg's formula is a simple reciprocal mathematical rearrangement of Balmer's formula:

$$\frac{1}{\lambda} = \frac{4}{B}\left(\frac{1}{4} - \frac{1}{n^2}\right), \quad \text{for } n = 3, 4, 5 \ldots$$

Rydberg then generalized his formula to take into account a new series of lines being observed for the hydrogen atom:

$$\frac{1}{\lambda} = \frac{4}{B}\left(\frac{1}{n_1^2} - \frac{1}{n_2^2}\right), \quad \text{for } n_1 = 1, 2, \ldots 6, \quad \text{and} \quad n_1 < n_2 < \infty$$

Today, $4/B$ is known as the Rydberg constant $R_H = 1.09737316 \times 10^7 \text{ m}^{-1}$, and is used in the formula to predict the complete spectrum produced by hydrogen. All of these series have been confirmed experimentally, as shown in Table 5.

BOHR'S SPARK OF GENIUS

Rydberg had been able to capture the observed periodicity of the hydrogen lines in an elegant formula, but he had no idea what underlying physical process could be responsible for such mathematical regularity.

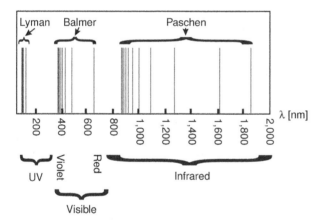

Figure 82   Spectral lines for the emission of hydrogen for $n_1 = 1$, $n_1 = 2$, and $n_1 = 3$. Looking at them on a graph shows the pattern generalized by Rydberg in his formula.

TABLE 5   Spectral Lines Predicted by Rydberg for the Emission of Hydrogen and Later Confirmed Experimentally

n_1	n_2	Name of series	Convergence
1	$2 \to \infty$	Lyman	91.13 nm (UV)
2	$3 \to \infty$	Balmer	364.51 nm (visible)
3	$4 \to \infty$	Paschen	820.14 nm (IR)
4	$5 \to \infty$	Brackett	1,458.03 nm (IR)
5	$6 \to \infty$	Pfund	2,278.17 nm (IR)
6	$7 \to \infty$	Humphreys	3,280.56 nm (IR)

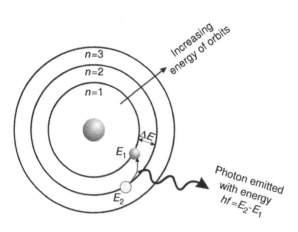

Figure 83   In the Bohr model of the hydrogen atom, electrons are allowed to circulate around the nucleus only at very specific orbits. Jumping from a higher orbit to a lower orbit causes the emission of a photon that carries the difference in energy between the two orbits.

Returning to 1911, we find the world of physics was faced with a major riddle about the atomic structure proposed by Rutherford. The three major questions were:

1. Why don't electrons fall into the nucleus?

2. Why don't atoms emit light in a continuous spectrum containing all colors?

3. Why do atoms emit light only at specific wavelengths unique to each element?

Enter Danish physicist Niels Bohr—Rutherford's new, young assistant at Manchester University. Bohr asked himself if the quantization that had been used by Planck to solve the blackbody radiation problem and by Einstein to explain the photoelectric effect was in fact a universal characteristic of energy. That is, maybe the emission or absorption of energy by any material object could only take place in discrete jumps, and never as a gradual, smooth change. If this were the case for Rutherford's atom, then very small changes in the electron's energy would be impossible, and the electron would remain trapped in its orbit.

Taking this idea further, Bohr boldly proposed that there are only certain stable orbits in which the electron can exist without crashing into the nucleus after spiraling to its death. Bohr restricted the allowed orbits to those in which the energy would be quantized into a value proportional to integer steps of h—the constant Planck had determined as the smallest quantum.[‡]

The assumption that only certain orbits would be allowed led Bohr to propose that an electron moving from one allowed orbit to another would either need to absorb or emit a certain quantized amount of energy.

So for the hydrogen atom shown in Figure 83, Bohr proposed that an electron would radiate a photon as it jumped from a higher orbit E_2 to a lower orbit E_1. The

[‡]Bohr's actual quantization condition is that the allowed angular momentum, L, of an orbiting electron (a measure of rotational motion of the electron) is given by $L = h/2m_e v$, where h is Planck's constant, m_e is the mass of the electron, and v its velocity.

photon would then carry an amount of energy equal to the loss of energy by the electron when jumping between these orbits:

$$E_{photon} = E_2 - E_1 = \Delta E$$

Remember that Einstein had discovered during his work on the photoelectric effect that the energy of a photon is equal to Planck's constant multiplied by the photon's frequency, so Bohr substituted $E_{photon} = hf$ into the prior equation to obtain:

$$\Delta E = hf$$

Bohr kept on exploring the math of his atom model, and through quite simple arithmetic and basic physics was able to arrive at Rydberg's constant just from the balance of energy lost by the electron when it jumped between allowed orbits! Remember that Rydberg arrived at his constant by fitting a formula to experimental data, but without any knowledge of the processes that generated the spectral lines. In contrast, Bohr arrived at the same constant from a purely theoretical basis, and demonstrated that Rydberg's constant can be derived from more fundamental constants:

$$R_H = \frac{e^4 m_e}{8\varepsilon_0^2 ch^3}$$

where

$\varepsilon_0 = 8.854 \times 10^{-12} A \cdot s/V \cdot m$ is the permittivity of free space, which is a constant that relates the mechanical force between two separated electric charges (in our case the electron and the nucleus) to the amount of charge they carry (in our case the absolute value of charge for the electron and the nucleus is the same),

$e = 1.602 \times 10^{-19}$ Coulomb is the charge of the electron (and the nucleus),

$m_e = 9.109 \times 10^{-31}$ kg is the mass of the electron,

$c = 299,792,458$ m/s is the speed of light, and

$h = 6.626 \times 10^{-34}$ J $\cdot$ s is Planck's constant.

Go ahead—take out your calculator and check if plugging the values of these fundamental constants results in the value that Rydberg found empirically. By the way, you may need to first simplify some of the powers of 10, since most calculators will round up h^3 to zero and won't let you make that division.

Bohr's achievement was astonishing. His model of the hydrogen atom behaved like the real thing, being able to predict the entire emission spectrum of the hydrogen atom based on fundamental physics plus the assumption of quantization! Bohr was later awarded the 1922 Nobel Prize in Physics "for his services in the investigation of the structure of atoms and of the radiation emanating from them."

ORBITALS AND NOT ORBITS

In spite of its success with explaining the spectral lines produced by the hydrogen atom, Bohr's atomic model is not perfect. It only works for atoms that have one electron. That is, it only applies to hydrogen, ionized helium (He^+), doubly ionized lithium (Li^{++}), and a few other light ions. More importantly, however, it does not justify Bohr's assumption that electrons moving in circular orbits would not lose energy by radiation. As we will see in the next chapter, it would take a decade before French physicist Louis de Broglie could explain the lack of radiation by proposing that the electron is not a classical particle traveling through an actual circular path. Today, we talk about the probability of an electron being in an *orbital*, rather than of an electron flying through space in an orbit.

Nevertheless, Bohr's basic concept remains valid. Atoms indeed radiate photons when electrons jump between an allowed higher energy level (Figure 84a) to a lower allowed energy level (Figure 84b). This is the basis for most of our modern solid-state electronics. For example, LEDs—those ubiquitous little indicator lights in all of our modern appliances—emit light because electrons in the atoms of their active materials jump from a higher energy level to a lower energy level. LEDs emit light of a very specific color that depends on the electron energy levels allowed by the atoms in the active material. One needs to supply the same amount of energy to the atom so the electron will jump from the lower energy level to the higher energy level.

A nice experimental proof of this is to determine Planck's constant h with some inexpensive LEDs of different colors[§] wired as shown in Figure 84c. At the very threshold for light emission, the voltage V_{LED} across the LED is just enough to provide energy to the electrons to jump between two energy levels. Therefore:

$$E = V_{LED} \times e = hf$$

where e is the electron charge ($e = 1.60 \times 10^{-19}$ Coulomb). To conduct the experiment, you will need to measure the voltage across each LED when light emission just starts, so do this in a darkened room. By the way, the emission from the IR LED is invisible to the human eye, but digital cameras and video cameras (especially those with "night-vision mode") are sensitive to IR, allowing you to see the IR LED light.

Once you have data for a number of LEDs, graph V_{LED} against f. If you can't find the LED's specs for the exact wavelength of the light emitted, use the values shown in Table 2. The slope of the graph you just created is equal to h/e, allowing you to roughly estimate the value of Planck's constant h. For reference, the recognized value for Planck's constant is $h = 6.626 \times 10^{-34}$ J · s.

Because of the dependence of the eye's threshold on color, and due to real-world properties of LEDs, this method of determining Planck's constant is not precise at all. However, it should give you a nice, instinctive feel for the dependence of photon energy (represented by its color) and the electron jump necessary to emit the photon.

[§]Make sure the LEDs are packaged in clear, not colored plastic.

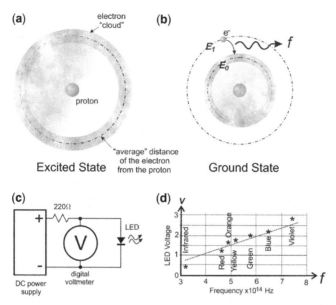

Figure 84   An atom in the excited state (**a**) decays to a stable "ground" level (**b**) by radiating energy. The absorption and radiation of energy takes place only in integer multiples of a fundamental unit of energy E, proportional through Planck's constant h to the photon's frequency f: $E = hf$. (**c**) A simple setup can be used to estimate Planck's constant h from the operating threshold of a number of LEDs that emit light at different wavelengths λ. (**d**) The slope of a linear fit of the data yields a rough estimate of h/e.

QUANTIZATION—THE CORE OF QUANTUM PHYSICS

Let's quickly review the major problems faced by classical physics at the turn of the twentieth century:

1. Classical physics would predict that a blackbody should radiate an infinite amount of UV light. However, hot objects actually produce a limited glow that peaks at a wavelength dependent on temperature and then decays to zero at short wavelengths.

2. Classical physics would predict that light waves hitting a metal surface should take time to build enough amplitude to shake electrons loose. In addition, the emission of electrons should be dependent on the intensity, and independent of the wavelength of the incident light. However, the photoelectric effect is instantaneous, independent of the intensity, but definitely dependent on wavelength.

3. Classical physics would predict that atoms shouldn't last because their electrons would spiral into the nucleus within an instant of their creation. However, everything around us is made of very stable atoms. In addition, classical physics would predict that the light produced by exciting atoms of a certain

element should contain all colors of the spectrum. However, the light produced by an element contains only discrete spectral lines that are unique to that element.

All of these problems were resolved in the same exact way. Namely, by using quantization involving Planck's constant $h = 6.626 \times 10^{-34}$ J · s:

1. In 1900, Max Planck resolved the Ultraviolet Catastrophe by assuming that the vibrations produced by a blackbody's harmonic oscillators are quantized by a new constant h that was later named in his honor.

2. In 1905, Albert Einstein solved the mystery of the photoelectric effect by proposing that light is not a wave, but rather a stream of quantized packets of energy (that we now call photons). The energy E of each quantum is given by its frequency f and Planck's constant h.

3. In 1913, Niels Bohr proposed to quantize the orbits of the hydrogen atom, thus explaining their stability and the discrete line spectra they emit. Bohr's quantization condition is that the allowed angular momentum L of an orbiting electron is given by $L = h/2m_e v$, where m_e is the mass of the electron, v its velocity, and h is again Planck's constant.

We can thus see the tremendous importance of Planck's constant as a fundamental characteristic of nature. The existence of Planck's constant tells us that the universe is not smooth at small scales, but is made of tiny bits. This is very much like the picture in your computer screen: although it looks like a smooth, continuous image, it is actually formed by tiny pixels.

Planck's constant is in essence a measure of the "graininess" of the universe at microscopic scales. It is very small—just 6.626×10^{-34} J · s, but not zero. If it were zero, then the universe would be perfectly continuous, and the predictions of classical physics would work out. However, h is not zero, and that makes the world of the microscopic behave in ways we don't expect at all based on our everyday experiences.

EXPERIMENTS AND QUESTIONS

1. Observe the light emitted by a hydrogen spectrum tube through a spectroscope. Can you distinguish the Balmer lines? How well do these lines fall at the positions predicted by Balmer's empirical formula $\lambda = B(m^2/m^2 - 4)$? For what values of n_1 and n_2 do your measurements match Rydberg's formula: $1/\lambda = 4/B(1/n_1^2 - 1/n_2^2)$, for $n_1 = 1, 2, \ldots 6$, and $n_1 < n_2 < \infty$?

2. Bohr calculated Rydberg's constant from fundamental constants using $R_H = e^4 m_e/8\varepsilon_0^2 ch^3$. How well does Bohr's calculation agree with Rydberg's empirically derived constant?

3. Using your spectroscope, observe the light emitted by an incandescent lightbulb, a fluorescent lightbulb, and a neon sign. What characterizes each spectrum? What is the main difference between the spectra from the fluorescent and neon tubes compared to the spectrum produced by the incandescent light? Explain.

4. What is unique about the spectrum obtained from a fluorescent light? What elements are used in fluorescent light fixtures?

5. Using your spectroscope, observe the light produced by a helium spectrum tube. What important observation was made by Rutherford based on the characteristic spectrum produced by helium?

6. How did Bohr modify Rutherford's atomic model to account for the discrete nature of atomic emission spectra? What are the shortcomings of Bohr's atomic model?

7. Why does each gas show a color only when electricity is passed through the discharge tube? What do the different colors in a line spectrum represent? Why do different substances show different spectra? Is it possible to identify the gases used in other light sources based on their emission spectra? How are astronomers able to identify chemicals in distant space?

8. Use the setup of Figure 84c to measure the threshold voltage at which LEDs of different colors just start to shine. Plot the threshold voltages against LED wavelength. Fit your measurements to a linear equation using linear regression. The slope of the graph you create is an estimate of h/e. Estimate Planck's constant h by assuming the charge of the electron $e = 1.60 \times 10^{-19}$ Coulomb. How well does your estimate of h match the accepted value for Planck's constant $h = 6.626 \times 10^{-34}$ J · s? What are possible sources of error? How could this experiment be improved?

9. Outline how quantization using Planck's constant h resolved some of the major problems faced by classical physics at the turn of the twentieth century.

CHAPTER 5

WAVE–PARTICLE DUALITY

In the last chapter, we saw the importance of the concept of quantization as a way of solving a large number of problems with which classical physics had struggled without success. Quantization certainly suggested that light—as Einstein had proposed in 1905—is a stream of particles. However, many physicists refused to take quantization as more than an elegant mathematical trick to solve the problems of blackbody radiation, photoelectric effect, and discrete emission lines.

An experiment conducted in 1923 at Washington University in St. Louis by American physicist Arthur Compton finally convinced all but the most die-hard physicists that light is indeed made out of particles. In Compton's experiment, light (in the form of X-rays) was made to interact with virtually free electrons.

It could be assumed that the electrons were free and at rest, so the solution to the problem shouldn't be bound by the special cases that needed the "trick" of quantization. Classical physics predicts that the electron should absorb energy from the light wave, and then re-emit the light at the same frequency. However, Compton's experiments actually showed that light bounces off the electron with lower energy, just as if the light were a stream of particles colliding with the electrons. That is, photons are able to transfer momentum to another particle. That is definitely the signature of a particle! Compton was awarded the 1927 Nobel Prize in Physics for his discovery.

Let's take a look at the geometry of Compton scattering as shown in Figure 85. A photon of frequency f collides with an electron at rest. Before the collision, the energy of the photon is given by Planck's formula $E = hf$. Upon collision, the photon bounces off the electron, giving up some of its energy, while the electron gains momentum. But the photon cannot lower its velocity, so the loss of energy by the photon shows up as a decrease in the photon's frequency f (or, conversely as an increase in its wavelength λ) since $E = hf$.

As we can see, the photon loses some energy after the collision, bouncing off at an angle θ with a new energy E' and momentum p', which translate into a decrease in frequency to f'. Of course, the initial photon's wavelength is λ, and its wavelength after the collision is λ'.

The photon's momentum is:

$$p_{photon} = \frac{E}{c} = \frac{hf}{c} = \frac{h}{\lambda}$$

Exploring Quantum Physics Through Hands-On Projects. By David Prutchi and Shanni R. Prutchi
© 2012 John Wiley & Sons, Inc. Published 2012 by John Wiley & Sons, Inc.

121

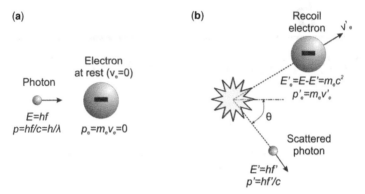

Figure 85   Compton Effect. **(a)** A photon with momentum $p_{\text{photon}} = E/c = hf/c = h/\lambda$ approaches an electron at rest. **(b)** After the collision, the photon transfers momentum to the electron, which results in a proportional decrease in the photon's frequency.

So if the change in the photon's wavelength is $\Delta\lambda = \lambda' - \lambda$, the electron gains the momentum lost by the photon:

$$\Delta\lambda = \lambda' - \lambda = \frac{h}{m_e c}(1 - \cos(\theta))$$

Please note that the wavelength shift (but not the frequency or energy shifts) is independent of the wavelength of the incident photon.

The maximum change in wavelength for the photon happens when it transfers as much momentum as possible to the electron. That is, when $\cos(\theta) = -1$. The maximum change in wavelength is thus:

$$\Delta\lambda = \frac{h}{m_e c}(1 - (-1)) = \frac{2h}{m_e c} = \frac{2 \times (6.63 \times 10^{-34}\,[\text{J} \cdot \text{s}])}{9.1 \times 10^{-31}\,[\text{kg}] \times 3 \times 10^8\,[\text{m/s}]}$$

$$= 4.86 \times 10^{-12}\,\text{m}$$

Even this maximum shift in wavelength would be insignificant for visible light with $\lambda \approx 10^{-7}$ m, but not for X-rays and gamma rays with $\lambda < 10^{-10}$ m. Please note that this is the maximum change, which doesn't mean that all the photons will recoil at $180°$ (to give $\cos(\theta) = -1$). In fact, most of the photons will bounce at much smaller values of θ.

GAMMA-RAY SPECTRUM ANALYSIS

To observe the Compton Effect we need a source of high-energy photons and a suitable spectrometer to observe the frequency shift in photons as they recoil. The source of short-wavelength photons is not much of a problem—a ^{137}Cs source produces photons at a wavelength $\lambda = 1.88 \times 10^{-12}$ m.

Actually, the more common unit for expressing the energy of high-frequency photons is in *electronvolts*. By definition, one electronvolt (eV) equals the amount of kinetic energy gained by a single electron when it accelerates through an electric potential difference of one volt. Thus, one electronvolt is equal to 1.602×10^{-19} J.

Since the energy of a photon using Planck's formula is $E = hf = hc/\lambda$, the energy of a photon in eV is: $E[\text{eV}] \approx 1{,}240[\text{eV} \cdot \text{nm}]/\lambda[\text{nm}]$. ^{137}Cs thus emits photons with 660 keV of energy, which is the common way in which you will see radioactive emission lines specified.

Spectrum analysis at these wavelengths is also relatively easy to do using a scintillation crystal coupled to a photomultiplier. Unlike the scintillation probe of Figure 61 however, we will need to couple the crystal to the PMT probe we built in chapter 2 (Figure 29). This is because the amplitude of the pulse output by a scintillator/PMT probe is proportional to the energy of the photon that causes the scintillation. Thus, as shown in Figure 86, spectral analysis with a scintillator/PMT is simply a matter of collecting a histogram of pulse heights. This histogram is essentially a representation of the number of photons detected by the scintillator that have been counted as a function of their energy.

The instrument used to histogram pulse heights is called a *multichannel analyzer* (MCA). It measures the energy spectrum and each channel, or *bin*, corresponds to a

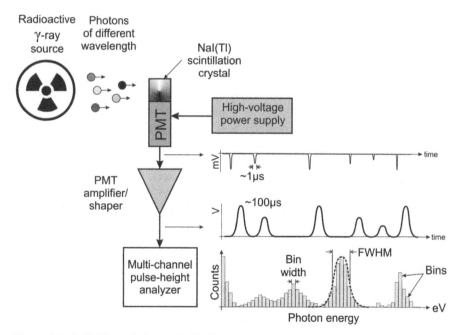

Figure 86   A PMT coupled to a scintillation crystal produces pulses of amplitude proportional to the energy of the gamma photons that interact with the crystal. These small, narrow pulses must be amplified and shaped so they can be processed by a multichannel pulse-height analyzer (commonly known as an MCA), which produces a histogram of pulse heights that represents the energy spectrum of incoming gamma photons.

narrow energy range. The histogram then represents the number of events within each bin, which is the energy spectrum.

Multichannel analyzers used to be expensive laboratory instruments. However, a very ingenious program was recently released for free by Marek Dolleiser of the University of Sydney in Australia, which allows a PC to act as an MCA. Pulse Recorder and Analyser (PRA)* analyzes scintillator signals input through the PC's sound card. It is easy to use, and performs amazingly well, considering that it performs real-time pulse recognition and display of spectral data. The program can also be used to count pulses from Geiger counters and other radiation detectors. Although it can't compete with the speed or performance of a commercial MCA as an analytical laboratory tool, this program is perfect for conducting our experiments!

PRA requires a PC running the 32-bit or 64-bit Windows® operating system. PRA can also run under Linux using Wine. The PC needs a sound card and very few other resources. We run it on an old Centrino® 1-GHz laptop with 256K RAM. We have found that older laptops and PCs with Windows XP (SP3) run best for this application, since some of the more recent machines micromanage the sound card to the point that PRA is rendered useless. If you have a commercial PMT pulse amplifier, you will probably have to stretch the output pulses so that they can be acquired through the sound card. We have an amplifier that outputs 2-μs pulses, and we stretch the pulses with a simple low-pass filter consisting of a 100-Ω resistor in series between the amplifier's output and the sound card input, and a 0.56-μF capacitor between the sound card's input and ground pins. PRA includes a good help file that you should read to get your system to work.

The choice of detector for gamma-ray spectrometry is important, because not all detectors are able to equally discriminate between photons that have only slight differences in energy. This quality is characterized by the so-called *energy resolution*, which is defined as the width (commonly specified as the "full width at half-maximum" or FWHM) of the spectrum for a spectral line at a certain energy. In other words, the energy resolution measures how much a single spectral line is smeared out by the detector (Figure 86, bottom graph).

A typical energy resolution for 662 keV gamma rays from ^{137}Cs of a small NaI(Tl) scintillator is around 8% FWHM. Unfortunately, for the same gamma photons, the resolution of a plastic scintillator is closer to 200% FWHM, so it would smear the spectrum so much we wouldn't be able to distinguish the 662-keV line from the Compton Shift. You will need a small (e.g., 1-in.-diameter × 1-in.-tall) NaI(Tl) scintillation crystal to perform this experiment.

Setting up the system to observe the Compton Effect is straightforward. Couple the NaI(Tl) scintillation crystal to the PMT's photosensitive face. Remember to use coupling grease to reduce losses, and make sure that no stray light enters the PMT. Then, simply connect the output of the PMT probe to the amplifier/discriminator that you built in chapter 2 (Figure 34). Power the PMT probe with a low-ripple, high-voltage power supply (e.g., Figure 31 or Figure 32), and hook up the "Analog

---

*Pulse Recorder and Analyser (PRA) is available for free download at the software author's website www. physics.usyd.edu.au/~marek/pra/index.html.

Output" connector of the amplifier/discriminator to the PC's microphone or line-in input (through either the right or left audio channel).

Place the scintillation probe and ^{137}Cs source facing each other, at least 30 cm above the work surface, and away from any solid objects. You should place the source far enough away from the detector so that the MCA (e.g., the PRA software) yields a clean spectrum, as shown in Figure 87a. You should be able to see the 662-keV line produced by the ^{137}Cs source, as well as the "Compton plateau," which is produced by Compton scattering of gamma rays within the NaI(Tl) scintillation crystal. Note that when the scattered gamma photon escapes from the crystal, only the energy deposited on the recoiling electron is detected. The upper edge of the plateau (the "Compton Edge") results from the most inelastic collisions—that is, those where the photon is scattered at an angle of 180°. You can use the equations that we saw before (page 122) for explaining the Compton Effect, and calculate the energy at the Compton edge from:

$$\frac{1}{E'} - \frac{1}{E} = \frac{[1 - \cos(180°)]}{m_e c^2} = \frac{2}{m_e c^2}$$

where E is the energy of the original gamma photon (662 keV for ^{137}Cs), and E' is the energy of the backscattered photon. The Compton edge thus sits at $E_e = E - E'$.

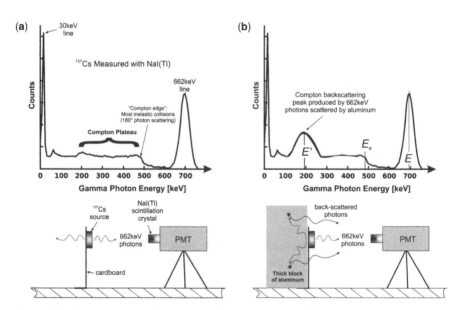

Figure 87   You can observe the spectrum of gamma photons being backscattered by the electrons in aluminum using this setup. (a) You should only see the 662-keV line produced by ^{137}Cs and some Compton-scattered photons within the NaI(Tl) crystal when the source and scintillation detector are placed far away from solid materials. (b) Placing a 5-cm-thick block of aluminum behind the source causes many photons to be backscattered through the Compton Effect.

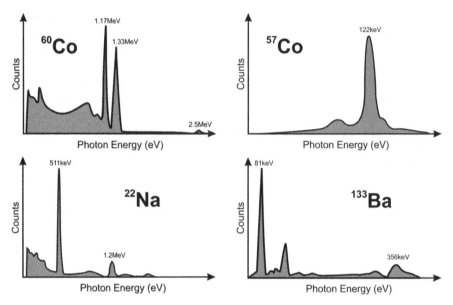

Figure 88   Stylized gamma-ray spectra for disk sources that you may use to test your NaI(Tl)/PMT probe and MCA, as well as to obtain further data points for E, E', and E_e.

Although that in itself is a demonstration of the Compton Effect, it is much more dramatic to place a thick aluminum block (at least 5-cm thick) behind the source, as shown in Figure 87b.[23] This will cause a large number of photons with energy E' to be backscattered by electrons in the aluminum. You can thus verify the Compton Effect by comparing the channel numbers at which the Compton peak (E') and Compton edge (E_e) appear in relationship to the channel number for the 662-keV ^{137}Cs line. If Dr. Compton was right to assume that photons are actual particles, then you should see the Compton edge E_e at around 480 keV, and the backscattered peak E' at around 180 keV.

The correctness of the Compton Effect can also be confirmed by calculating the mass of the electron m_e from the measurement of the Compton edge and the Compton peak[24]:

$$\frac{1}{E'} - \frac{1}{E} = \frac{[1 - \cos(180°)]}{m_e c^2} = \frac{2}{m_e c^2}$$

Using your measurements of E, E', and E_e for a ^{137}Cs source and an aluminum block, estimate the rest mass of the electron. You may also want to use other gamma-ray sources (Figure 88) to obtain more data points to refine your estimate of m_e by measuring E, E', and E_e for each. How well does your estimate agree with the accepted value of $m_e = 9.11 \times 10^{-31}$ kg or its equivalent of 511 keV?

WHAT IS THE NATURE OF LIGHT?

So light is a particle, right? But wait! What about diffraction and interference? Didn't Foucault show that the speed of light in air and water needed to explain diffraction disagree with experimental data if we assume that light is a stream of particles (chapter 1, Figure 7)? And isn't interference supposed to be the obvious signature of a wave? So, what is the nature of light? Is light a wave or a flow of particles?

In trying to solve this quandary, it is interesting to think what would happen if we combined the double-slit experiment (chapter 1, Figure 4) with the single-photon experiment (chapter 2, Figure 33). American physicist Richard Feynman, who won the 1965 Nobel Prize in Physics, liked to introduce this hybrid experiment by thinking about what would happen if a machine gun is shot at an iron plate with two slits in it (Figure 89a). If there were a concrete wall behind the iron plate, and since bullets don't

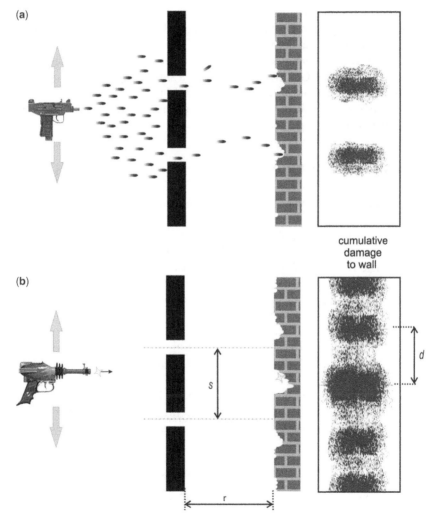

Figure 89   In Feynman's thought experiment, a machine gun shoots bullets—one at a time—against a steel plate with two slits. (**a**) The damage to the brick wall behind it is expected to be behind these slots. (**b**) When the experiment is conducted with photons shot one at a time, their cumulative hits actually form an interference pattern.

interfere with each other like water waves, the bullets would be expected to carve the brick wall behind the two slots. A few bullets may bounce off the edges of the slots, creating just a few stray hits, but the vast majority of the damage to the wall would be expected to be in the two areas behind the slots.

Surprisingly, when this experiment is carried out with single photons (assuming the wall would record the individual photon hits, as would happen if it were a photographic plate), the cumulative pattern is not the one predicted in the thought experiment of Figure 89a. Instead, the individual photon hits accumulate to form an interference pattern, as shown in Figure 89b. This result could be interpreted as meaning that light really is conveyed as a wave, since interference occurs, but at the same time, a single hit is detected on the screen at any given time, indicating that single photons are flying through the apparatus. Hence, the emergence of an interference pattern suggests that each photon interferes with itself, and therefore each photon must be going through both slits at once!

In 1909, about a century after Young carried out his original double-slit experiment, British physicist G. I. Taylor performed a similar experiment in a very dark room, using a photographic plate as the screen on which he showed that even the feeblest light source—equivalent to "a candle burning at a distance slightly exceeding a mile"—could lead to interference fringes.

TWO-SLIT INTERFERENCE WITH SINGLE PHOTONS

Taylor exposed photographic plates for up to 3 months to obtain interference patterns using his very weak light source. Today, we can conduct the same experiment within a few minutes, using the setup shown in the block diagram of Figure 90. The basic idea remains the same—to illuminate the double slit with a very weak beam. So weak in fact, that there is a reasonably high assurance that only a single photon of the laser beam is present in the device at a time.

As shown in Figure 90, the distance between our two-slit slide and the detector is 10 cm, which means that a photon takes:

$$\text{Time} = \frac{\text{Distance}}{\text{Velocity}} = \frac{0.1 \ [\text{m}]}{300,000,000 \ [\text{m/s}]} = 0.33 \ \text{ns}$$

to traverse this distance. As such, at most one photon hits the photocathode as long as photons arrive at a frequency of less than $1/0.33 \ [\text{ns}] = 3.03 \times 10^9 [\text{photons/s}]$.

Our HeNe laser produces photons at a wavelength of $\lambda = 632$ nm, each with an energy:

$$E = hf = h\left(\frac{c}{\lambda}\right) = 6.626 \times 10^{-34} \ [\text{J} \cdot \text{s}] \times \left(\frac{2.998 \times 10^8 \ \left[\frac{\text{m}}{\text{s}}\right]}{632 \times 10^{-9} \ [\text{m}]}\right) = 3.143 \times 10^{-19} \ \text{J}$$

We measured the optical power output of our laser to be 5 mW, so we can calculate the number of photons produced by our laser each second, just as we did in

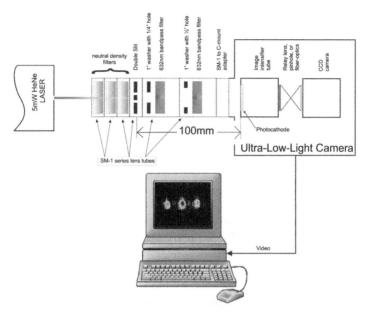

Figure 90   This is the experimental setup that we constructed to demonstrate single-photon interference. Just as in the experiment of Figure 33, a laser beam is attenuated with neutral-density filters of at least $D = 6.73$, to the point where only one photon is present within the apparatus. Each photon then passes through a double slit. We used an extremely sensitive TV camera to detect the individual photons. Two HeNe-line band-pass filters (Edmund Industrial Optics model NT43-133) and two apertures (1-in. flat washers from the hardware store) keep stray, nonlaser photons from adding to the noise.

chapter 2 (Figure 33):

$$\frac{5\,[\text{mW}]}{3.143 \times 10^{-19}\left[\dfrac{\text{J}}{\text{photon}}\right]} = \frac{5 \times 10^{-3}\left[\dfrac{\text{J}}{\text{s}}\right]}{3.143 \times 10^{-19}\left[\dfrac{\text{J}}{\text{photon}}\right]} = 1.591 \times 10^{16}\left[\frac{\text{photons}}{\text{s}}\right]$$

By placing the attenuators before the two-slit slide, we can be assured there will be at most one photon within the 10-cm path between the slits and the detector if the density of the attenuators is at least:

$$D = \log_{10}\left(\frac{1.591 \times 10^{16}\left[\dfrac{\text{photons}}{\text{s}}\right]}{3.03 \times 10^{9}\left[\dfrac{\text{photons}}{\text{s}}\right]}\right) = 6.72$$

Therefore, any density for the system over $D = 6.72$ results in an average population of less than one photon at a time within the optical assembly.

As we saw before, the optical densities of neutral-density filters used in combination are additive. We use a combination of three $D = 1$ (Thorlabs NE10A), one

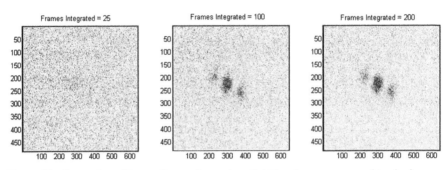

Figure 91   Integration of image frames from a low-light-level camera exposed to single photons flying through a double-slit apparatus. The interference pattern emerges slowly from the background noise. The camera that we used has an onboard integrator, so each frame integrated by the computer is the result of 64 frames integrated by the camera itself. In turn, each frame is exposed for up to 33 ms.

$D = 2$ (Thorlabs NE20A), and a $D = 5$ (Thorlabs NE50A) filter for a total of $D = 10$, allowing a maximum of 1.6×10^6 photons/s to reach the double slit.

The attenuated beam illuminates a 45 μm-separation double slit (cut out from an IF-508 diffraction mosaic slide from Industrial Fiber-optics). Since our HeNe laser produces a 0.8-mm-diameter spot with a Gaussian distribution, only about 7% of the photons can actually pass through the narrow slits toward the detectors. Each of the laser-line filters allows only 45% of desirable 632-nm photons through, so only 22,680 photons/s reach the detector.

The average time between photon arrivals at the camera is thus 44 μs. Since any individual photon goes across the 10-cm length of the apparatus with a flight time of 0.33 ns, there is on average a photon in flight between the neutral-density filter and the camera for only about 0.33 ns out of every 44 μs. Thinking in particle terms, this means that for 99.99925% of the time there are "no" photons in flight, while for only 0.00075% of the time there is "one" photon in flight in the apparatus. Since photon bunching is insignificant, the probability of there being two photons in flight simultaneously is thus negligible. All the interference effects observed in this apparatus can thus be attributed to the "weird" behavior of individual photons, traversing through the double slit one at a time!

In this case, the interference pattern doesn't appear instantaneously, as it does when the double slit is illuminated by the laser beam without attenuation. Since only a single photon strikes the camera's sensor at a time, only localized impacts of individual photons are observed. The interference pattern appears only after adding up (*integrating*) many of these individual, localized impacts.

We integrated successive video frames using the short MATLAB program shown in Listing 1. The freeware utility VFM[†] (Vision for MATLAB by Farzad

---

[†]Vision for MATLAB by Farzad Pezeshkpour is available for free download from www.mathworks.com/matlabcentral/fileexchange/authors/1939. An alternative for other versions of windows and Matlab is VCAPG2 by Kazuyuk: Kobayash: available at http://www.ikko.k.hosei.ac.jp/∼matlab/matkatuyo/vcapg2.htm. For non-MATLAB users, a number of freeware and shareware programs that integrate image frames on a PC are available for video astronomy applications. Examples of these are ASTROSNAP (www.astrosnap.com) by Axel Canicio and REGISTAX (www.astronomie.be/registax) by Cor Berrevoets.

Pezeshkpour) allows MATLAB to access video source devices, such as frame-grabber boards and USB cameras. Our program first captures the static dark background image that almost every camera produces with no light impinging on it. Frames are acquired and added to each other, one at a time. The dark static image is subtracted from each frame during the integration process to make it easier to distinguish the strikes produced by the photons. As a result, the interference pattern emerges slowly from the background noise as shown in Figure 91.

Listing 1   MATLAB program to integrate image frames.

```
%----------------------------------------------------------------
% Matlab program to integrate video image frames from single-
photon
% interference setup.
%
% © David Prutchi, Ph.D. 2008
%
% This program uses VFM (Vision for Matlab) to acquire
% image frames through a Windows video capture card.
%
%----------------------------------------------------------------
% From the VFM preview window, configure VFM to 640x480 video
resolution
%
noise(480,640) = 0; % initialize array that will hold static
dark image
a = vfm('grab',10); % acquire 10 frames
dark = double(a); % convert from image format to double-
precision numbers
% Integrate 10 frames
for n = 1:10
    noise =
noise+dark(:,:,1,n)+dark(:,:,2,n)+dark(:,:,3,n);
end
noise = noise/10; % Calculate average dark static image
clf        % Clear figure
imagesc(noise)     % Display average dark static image
colorbar
title('Dark Static Image')
drawnow
clear a       % Clear the image buffer
clear dark    % Clear the double-precision image buffer
% Ask user to turn LASER on
beep
disp('PLEASE TURN LASER ON')
pause(2)
```

```
beep
pause(2)
beep
pause(2)
c(480,640) = 0; % initialize array that will hold
interference pattern
for n = 1:200   % number of frames to be integrated
  a = vfm('grab',1);
  b = double(a);
  n
  c = c+b(:,:,1)+b(:,:,2)+b(:,:,3)-noise;
 if n = = 1
  figure(2)
  subplot(2,3,1)
  imagesc(c,[0,max(max(c))])
  title('Frames Integrated = 1')
  drawnow
 end
 if n = = 10
  figure(2)
  subplot(2,3,2)
  imagesc(c,[0,max(max(c))])
  title('Frames Integrated = 10')
  drawnow
 end
 if n = = 25
  figure(2)
  subplot(2,3,3)
  imagesc(c,[0,max(max(c))])
  title('Frames Integrated = 25')
  drawnow
 end
 if n = = 50
  figure(2)
  subplot(2,3,4)
  imagesc(c,[0,max(max(c))])
  title('Frames Integrated = 50')
  drawnow
 end
 if n = = 100
  figure(2)
  subplot(2,3,5)
  imagesc(c,[0,max(max(c))])
  title('Frames Integrated = 100')
  drawnow
 end
```

```
if n = = 200
 figure(2)
 subplot(2,3,6)
 imagesc(c,[0,max(max(c))])
 title('Frames Integrated = 200')
 drawnow
end
figure(3)
clf
imagesc(c,[0,max(max(c))])
colorbar
drawnow
end
```

IMAGING SINGLE PHOTONS

A regular TV camera wouldn't be able to detect anything at the low-photon flux we need to ensure only one photon passes the slits at a time. An "image-intensifier tube"—like those used by soldiers to see at night—is needed to make the image visible to a conventional camera element (e.g., a CCD camera).

In our single-photon experimental system (Figure 90), the photons that pass through the double slit strike the photocathode of one of these image-intensifier tubes. Just as with a PMT, these photons cause the release of electrons via the photoelectric effect, and are multiplied and accelerated within the intensifier tube. However, instead of finally hitting an anode, the secondary electrons are focused on a phosphor screen. Each incident photon that strikes the photocathode surface causes the release of many photons from the fluorescent screen, making it possible for a standard video camera to record the weak flux of photons passing through the double slit.

The gain and noise of the intensifier tube determine how low a photon flux can be reached and still yield an image of the interference pattern. So-called "Generation I" image-intensifier tubes are simple in design. As shown in Figure 92a, they utilize only a single potential difference to accelerate electrons from the photocathode to the anode (screen). Therefore, they achieve only moderate gain (a few hundred times), but provide high image resolution, a wide dynamic range, and low noise. In contrast, Generation II and Generation III tubes employ true electron multipliers to boost gain. That is, not only the energy but also the number of electrons between the photocathode and the screen is significantly increased. Multiplication is achieved by use of very thin plates of conductive glass containing scores of minute holes (each a few micrometers in diameter), inside which a cascade of secondary electron emission occurs. As shown in Figure 92, each hole in a *microchannel plate* (MCP) acts as a miniature PMT! The difference between the second and third generations resides in the type of material used in the photocathode. The "multialkali" photocathodes in second generation tubes yield a current of around 300 μA per lumen, while the gallium-arsenide photocathodes in third generation tubes have a luminous sensitivity

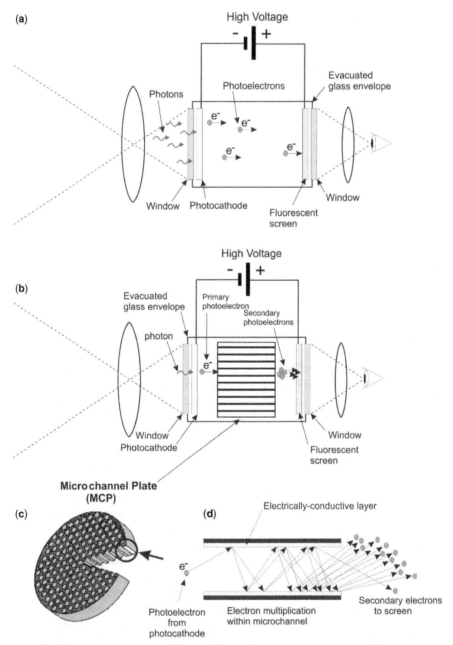

Figure 92   Each incident photon that strikes the photocathode surface of an image-intensifier tube causes the release of many photons from the fluorescent screen. (**a**) Generation I tubes utilize a potential difference to accelerate electrons from the photocathode to the anode (screen) to achieve moderate gains. (**b**) Generation II and III tubes employ MCP electron multipliers to boost gain. (**c**) MCPs are very thin plates of conductive glass containing scores of micrometer-sized diameter holes, in each of which (**d**) a cascade of secondary electron emission occurs, thus acting as a miniature PMT.

of approximately 1,200 µA per lumen, meaning that "Gen III" intensifier tubes can reach higher gains (up to 10^7 photons from the screen for every photon hitting the photocathode) than Gen IIs. Gain doesn't come for free though—in general, Gen II and Gen III intensifier tubes have lower resolution and produce more noise than Gen I tubes.

There are a number of options when acquiring or building a camera with sufficient sensitivity. The easiest way is to purchase a camera with a built-in intensifier, commonly used in fluorescence microscopy, astronomy, nighttime surveillance, and other applications that require image acquisition under very-low-level illumination. Depending on the technology, these go by the name of "ISIT Camera" (intensified silicon-intensified target), "IST Camera" (intensified silicon target), "ICCD" (intensified CCD), and "Photon-Counting Camera." They are expensive—upward of $8,000. However, they can often be found for just a few hundred dollars on eBay and at surplus laboratory supply houses.

We opted instead for building our own intensified camera using components that we purchased on the surplus market. Our setup, shown in Figure 93, uses a surplus Gen III image-intensifier tube (an MX-10160 Gen III intensifier tube used in the helmet-mounted AN/AVS-6 "ANVIS" aviation night vision imaging system, which we purchased on eBay) followed by a pinhole-lens black-and-white CCD camera. The small plastic box contains two AA cells to supply 3 V to the intensifier tube. For convenience, we built our optical tube using Thorlabs' SM1 and SM2 optical tubes and components. We fabricated the housing for the image-intensifier tube and light shields from surplus telescope-to-camera adapter tubes. The CCD camera looks directly at the intensifier's phosphor screen. We shielded all potential photon leakage gaps with thick black electrician's tape.

We found that the performance of our Gen III intensifier is superb, since it allows us to use $D > 10$ attenuators and see the interference pattern build up one photon at a time. With this setup, one can truly see single photons arriving to the detector, slowly building up the interference pattern from seemingly random hits. Watching this as it happens is a very moving experience for anyone who understands that under mundane, commonsense rules we should see two lines of light aligned with the slits. Instead, somehow, the single photons interfere with themselves! That means that each photon behaves as if it somehow flies through both slits at the same time!

THE ANSWER: COMPLEMENTARITY

So, yet again, what is the answer? Is light a wave, or is light a stream of particles? Well, actually it's neither (or both). Light apparently is something different altogether, but it *behaves* as a wave when the experiment is designed to reveal its wave-like properties, while it *behaves* as a particle when the experiment is designed to show its particle-like properties. This schizophrenic personality of light is known as the *"wave–particle duality."*

The type of experiment will show either light's particle-like or wave-like behavior, but not both at once, which made Niels Bohr state that the wave and particle

Figure 93 In our single-photon interference setup, a black-and-white CCD camera looks directly at the phosphor screen of a surplus Gen III image-intensifier tube. The interference pattern is projected directly onto the intensifier tube's photocathode.

aspects of light are complementary to each other. The concept of *complementarity* derives directly from wave–particle duality, and states that all physical reality is determined and defined by manifestations of properties that are limited by trade-offs between complementary pairs of these properties.

MATTER WAVES

Even in the light of the Compton Effect, critics of the early single-photon interference experiments dismissed the importance of the observation by noting that a photon doesn't have mass. Through some fancy hand-waving, they argued that the low-light interference could be caused through splitting and recombining the light quanta's wavefront. Decisive proof would come when particles with mass would show interference.

In 1924, French physicist Prince Louis-Victor de Broglie (pronounced "de Broy," and by the way, he did belong to the French royalty) proposed that maybe it's not just light that has this dual personality, maybe it's everything! He reasoned that if the quanta of light could be both a wave and a particle, then maybe the same could be true of electrons.

Remember that although a photon doesn't have mass, it does have momentum p given by:

$$p_{photon} = \frac{E_{photon}}{c} = \frac{hf}{c} = \frac{h\left(\frac{c}{\lambda}\right)}{c} = \frac{h}{\lambda} \quad \Rightarrow \quad \lambda = \frac{h}{p_{photon}}$$

De Broglie's hypothesis was that matter—which is commonly described by our perception as "solid"—can also behave as a wave. He took the concept used to find the momentum of a photon, and applied it to particles, proposing that the wavelength associated with a particle is:

$$\lambda = \frac{h}{p} = \frac{h}{mv}$$

where the momentum p of a particle equals the product of its mass m, and velocity v. In the same way, he used Planck's relationship to propose that the frequency f of the matter wave is related to the energy E of the particle by:

$$f = \frac{E}{h}$$

MATTER WAVES AND THE BOHR ATOM

As you may remember from chapter 4, Bohr was able to explain the discrete spectral lines emitted by the hydrogen atom by forcing the electrons into a limited number of permitted orbits (Figure 94a). However, like Planck before him, he did this without having a physical justification.

The first theoretical success of de Broglie's matter waves came when he used them to explain why certain electron orbits are allowed, and why others are not. He proposed that the allowed orbits for an electron are those that support a standing wave of specific wavelength, energy, and frequency (i.e., Bohr's energy levels); much like a guitar string sets up a standing wave when plucked.

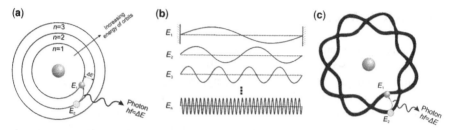

Figure 94   De Broglie used his proposed matter waves to explain why Bohr's atomic orbits would be quantized. **(a)** Bohr explained the discrete spectral lines of hydrogen by enforcing a limited number of orbits. However, he did this without having a physical justification. **(b)** De Broglie proposed that, like in a guitar string fastened at the ends to rigid supports, allowed orbits would be those where a complete number of electron waves would fit on the orbit. **(c)** A higher-energy orbit would thus fit a larger number of complete vibrations than a lower-energy orbit.

As shown in Figure 94b, de Broglie proposed that if one were to straighten an orbit into a string and fix the ends to rigid supports, it could be set to vibrate. However, since the ends are fixed, the only vibrations that it can support are those where full wavelengths fit between the ends. De Broglie proposed that these vibrations would correspond to his matter waves, so a higher frequency vibration would correspond to an electron at a higher energy state. As shown in Figure 94c, de Broglie's view of Bohr's atom consisted of electron waves of different wavelengths. An electron wave would transition between these wavelengths (each at one of Bohr's energy levels), giving off or absorbing photons with an energy hf equal to the difference in energy ΔE between the electron waves.

EXPERIMENTAL CONFIRMATION OF DE BROGLIE'S MATTER WAVES

Experimental confirmation of de Broglie's formula came in 1927, when G. P. Thomson at the University of Aberdeen and C. J. Davisson with L. H. Germer at Bell Labs observed diffraction—a typical wave-like behavior—from an electron beam.

Unlike photons, electrons have a rest mass, and are thus perceived as "solid" particles. Electrons are negatively charged and can be accelerated with ease, as is commonly done inside a CRT. Let's suppose that electrons, which have a mass of $m = 9.1 \times 10^{-31}$ kg and charge of $e = 1.60 \times 10^{-19}$ Coulomb are fired by an electron gun operated at a potential of $V = 4\,\text{kV}$ (Figure 95a). Their kinetic energy is:

$$E_{\text{kinetic}} = eV = 1.6 \times 10^{-19}\,\text{Coulomb} \times 4{,}000\,\text{V} = 6.4 \times 10^{-16}\,\text{J}$$

Since

$$E_{\text{kinetic}} = eV = \frac{mv^2}{2} = \frac{(mv)^2}{2m} = \frac{p^2}{2m}$$

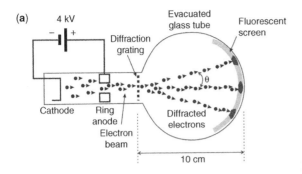

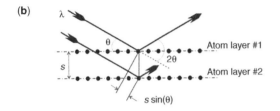

Figure 95 Electron diffraction can be observed in the laboratory with a specialized CRT. (a) Simplified representation of a vacuum tube used to observe the wave-like nature of electrons. (b) Simplified view of diffraction of electrons by a crystal.

the momentum of these electrons accelerated at 4 kV is thus:

$$p = \sqrt{2mE_{\text{kinetic}}} = \sqrt{2 \times 9.1 \times 10^{-31} \, [\text{kg}] \times 6.4 \times 10^{-16} \, [\text{J}]}$$
$$= 3.41 \times 10^{-23} \left[\frac{\text{kg} \cdot \text{m}}{\text{s}}\right]$$

which according to de Broglie gives them a wavelength of:

$$\lambda = \frac{h}{p} = \frac{6.626 \times 10^{-34} \left[\dfrac{\text{kg} \cdot \text{m}^2}{\text{s}}\right]}{3.41 \times 10^{-23} \left[\dfrac{\text{kg} \cdot \text{m}}{\text{s}}\right]} = 1.94 \times 10^{-11} \, \text{m} = 0.019 \, \text{nm}$$

De Broglie's hypothesis would prove correct if one could shoot electrons at a double slit and obtain an interference pattern.

Let's see how we could do that in the lab. Look at Figure 95a, and remember that when we experimented with microwave two-slit interference (Figure 18), we found that fringes appear at angles that satisfy $s(\sin \theta) = \lambda$. This formula also applies to simple ruled diffraction gratings, where the condition for diffraction is $\lambda = s(\sin \theta)$, or $\theta = \lambda/s$ for small angles. Since the experiment has to be carried out in a vacuum tube, we could assume a 10-cm-diameter bulb. Therefore, the grating spacing between rulers to produce interference at 1 cm would be:

$$s = \frac{\lambda}{\sin \theta} = \frac{0.019 \, [\text{nm}]}{0.1} \approx 0.2 \, \text{nm}$$

Although producing an artificial grating with rulings of approximately $s = 0.2$ nm is no easy task, the lattice of crystals can act as natural "gratings." Graphite was used by the Bell Labs team in 1927 to observe electron diffraction. And for a very good reason—a piece of graphite can be assumed to have its carbon atoms organized in a simple cubic lattice with the right spacing: 1 mol of carbon atoms (remember Avogadro's number $= 6.022 \times 10^{23}$?) weighs 12 g, and since the density of carbon is approximately 2 g/cm^3, 1 cm^3 contains $\sim 10^{23}$ atoms spaced 0.215 nm apart.

Davisson and Germer fired an electron beam at a thin graphite crystal, and observed that the beam was diffracted according to the angles given by the formula usually used for the diffraction of light (Figure 95b):

$$2s(\sin \theta) = n\lambda, \ (n = 1, 2, 3, \ldots)$$

This clearly showed that de Broglie was right. It's not only that light sometimes acts as particles, but particles with mass sometimes act as waves. Particle–wave duality seems to be a universal characteristic of everything that exists. The 1929 Nobel Prize in Physics was awarded to Louis de Broglie "for his discovery of the wave nature of electrons."

There are a number of commercially available tubes that are specifically made for the didactic demonstration of electron diffraction. They are made by Leybold (model 555626), PHYWE Systeme (model 06721.00), and Teltron (models 555 and 2555). We purchased one of the Teltron 555 tubes on eBay for around $100 (a new 2555 sold by Tel-Atomic is over $850) and hooked it up as shown in the diagram of Figure 96a. The source of electrons in the 555 tube is an indirectly heated cathode. The heater is a filament that needs to be supplied from a 6.3-V transformer designed specifically to supply vacuum tube filaments, for example a model 166F6 by Hammond Manufacturing. Once the cathode temperature has been allowed to stabilize, the voltage between the cathode and the anode is increased, so electrons are focused and accelerated by the cylindrical portion of the anode. A micro-mesh nickel grid coated with graphite is the target for the converging beam of electrons. The graphite coat is only a few molecular layers thick and can easily be damaged by a current overload. For this reason, the anode current should be monitored and never allowed to exceed 200 μA. An internal resistor biases the cathode, but an external 0–50 VDC power supply is used to focus the beam of electrons, which sharpens the diffraction pattern and allows better observation, especially at lower cathode-to-anode voltages.

If graphite had a simple cubical structure, we would expect to see four spots on the screen. As shown in Figure 96b, however, graphite has a hexagonal structure; because of this sixfold symmetry, it should yield six spots on the screen for s_1 and six for s_2. Each of the six spots for each s should cause the same amount of diffraction but should rotate about the direct beam, each separated by 60°. However, the hexagonal structure is made of planes of hexagonal rings only loosely bonded to each other, resulting in many minicrystal planes randomly oriented to one other. The resulting first-order diffraction pattern from this randomly oriented powdery crystal is thus made of two rings, one for each spacing s.

As shown in Figure 96c, two prominent rings are observed on the screen at a voltage of 4 kV. For simple diffraction, the condition is $\lambda = s(\sin \theta)$, which for

(a)

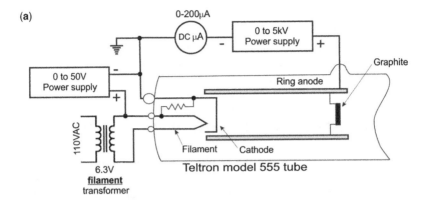

(b)

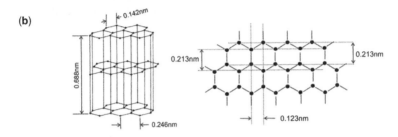

(c)

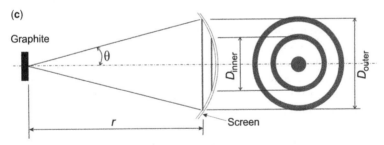

Figure 96   Geometry to interpret the diffraction-tube results. (**a**) Connection of Teltron model 555 electron-diffraction tube. (**b**) Graphite has a hexagonal crystal lattice structure, in which the planes on the right are responsible for the first two rings of the diffraction pattern. (**c**) The diameter of the rings depend on the diffraction angle and geometry of the tube.

small angles may be simplified as $\lambda = s\theta$. The Teltron 555 tube has a target-to-screen distance r of 14 cm, so the expected diameters for the circles can be calculated as:

$$D\frac{s}{2r} = \lambda \quad \Rightarrow \quad D = \frac{0.28[\text{m}] \times \lambda}{s}$$

Measured diffraction ring diameters need to be compensated for the curvature (66 mm) and thickness (1.5 mm) of the 555 tube. When this is done, and the voltage is varied between 2.5 and 5 kV, one can see that the tube produces diffraction ring diameters that accurately match the predictions made by de Broglie's hypothesis.

Build one of these setups and try it out yourself! This should be a deeply moving experience if you understand its implication: a particle of matter is not a "solid ball" but can behave as an ethereal wave!

TWO-SLIT INTERFERENCE WITH SINGLE ELECTRONS

De Broglie's critics argued that the results from electron-diffraction experiments may indicate an undulating interaction between electrons, so these experiments didn't decisively prove that electrons *are* waves when studied by diffraction. Definitive proof would still have to wait until technology advanced to make it possible to conduct the double-slit experiment with individual electrons. In one of his famous lectures, Feynman explained the ultimate significance of observing single-electron interference:

> We choose to examine a phenomenon which is impossible, absolutely impossible, to explain in any classical way, and which has in it the heart of quantum mechanics. In reality, it contains the only mystery.

The first double-slit experiment with single electrons was performed by Italian physicists Pier Giorgio Merli, GianFranco Missiroli, and Giulio Pozzi[25] in Bologna in 1974. However, their electron source (a thermionic emitter) wasn't sufficiently stable to absolutely ensure that only single electrons would arrive at the detector at the same time. However, in 1989, a group at Hitachi Labs led by Japanese physicist Akira Tonomura developed a double-slit system that allowed them to observe the build-up of the fringe pattern with a very weak electron source.[26]

Figure 97 shows a schematic representation of the modifications that Tonomura made to a transmission electron microscope (TEM) to develop his experimental setup. Electrons are emitted from a very sharp tungsten tip when a potential difference of 3–5 kV is applied between the tip and a first anode ring; this effect is known as "field emission." These electrons are then accelerated to the second anode potential of 50 kV (the de Broglie wavelength for the accelerated electrons is $\lambda = 0.0055$ nm). Assorted "electron optics" within the modified electron microscope attenuate and focus the electron beam so that a current of barely 1.60×10^{-16} A (that is only 1,000 electrons/s) is beamed toward the double slit.

The double slit is actually an extremely fine wire filament (1-μm diameter) placed between two conductive plates set 1 cm apart. The wire is biased at a positive voltage of 10 V relative to the plates. This arrangement is known as an *electron biprism*. Despite its complicated name, you should find this familiar: it is the electron equivalent of the interference experiment we conducted using a thin hair (Figure 4c).

Obviously, any electrons that make it past the biprism must have gone either through one or the other side of the fine wire. The discussion of the interference pattern generated by the electron biprism is beyond the scope of this book, but suffice it to say that if electrons act as waves, they should produce fringes with a spacing of 700 nm for the de Broglie wavelength in this setup.

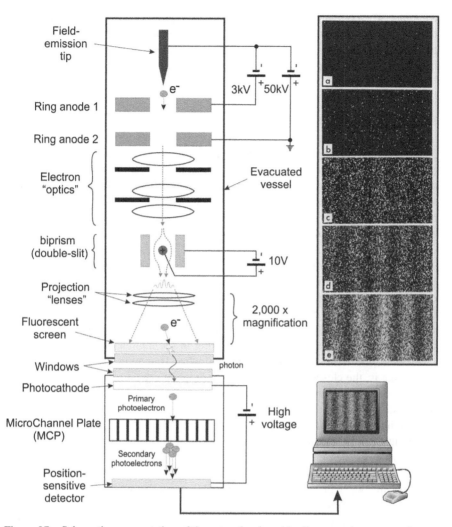

Figure 97　Schematic representation of the setup developed by Tonomura's group to observe single-electron interference. Inset (**a–e**): results of a double-slit experiment performed by Tonomura showing the build-up of an interference pattern of single electrons. Integrated number of electrons detected by the detector are (**a**) 10, (**b**) 200, (**c**) 6,000, (**d**) 40,000, and (**e**) 140,000. The inset was obtained from Wikimedia Commons at http://commons.wikimedia.org/wiki/Image:Double-slit_experiment_results_Tanamura_2.jpg. It was released to the public domain with kind permission of Dr. Tonomura. This insert is under the terms of the GNU Free Documentation License.

Two electron lenses then magnify the interference pattern 2,000 times and project it onto a fluorescent screen. Each 50-keV electron hitting the screen produces about 500 photons that generate photoelectrons inside an intensified position detector. This device works in a manner similar to an image intensifier, the only difference being that it produces an x,y coordinate for each hit, which can be read directly by a computer. The computer then integrates the hits to produce a final image of the electron interference pattern as shown in the inset of Figure 97. Through which slit did each of the electrons go? The answer is the same as before: somehow each electron behaves as if it is going through both slits at the same time!

A SIMPLE TEM

Systems to observe single-electron interference have been built by other groups, and a committed experimenter[‡] who would like to build such a device should study the excellent papers written by these research groups. This will help you to gain an understanding of how they overcame some of the difficult technical challenges involved. Even with today's technology, a system to conduct a single-electron interference experiment is a difficult project that should be undertaken only by very advanced experimenters.

In spite of this, one can work with the basic electron "optics" components to understand the operation and significance of Tonomura's experiment. Moreover, one can perform the basic experiments that demonstrate complementarity for the electron by accepting that in these experiments individual electrons behave in the same way as a beam consisting of millions of particles at a time. Let's consider this for a moment—can you see that what is happening in the humble Teltron 555 electron-diffraction tube is in principle the same as what Tonomura did one electron at a time? That is, the closely spaced atoms of a graphite crystal serve as the slit system, and the electrons exhibit the same interference effects as the photons in an optical double slit.

With this in mind, let's take a look at the construction of a basic electron optics system that can be used to experiment with different aspects of the electron's complementary properties.

Figure 98 shows our simple electron microscope. It is identical to the cold-cathode CRT that we built in chapter 3 (Figure 41), except that we added a sample within the anode's exit orifice and a large coil that acts as an electron lens. We evacuate the tube with the refrigeration-service vacuum pump (Figure 35), and power the tube with the same high-voltage power supply we built to energize our CRTs (Figure 39).

The "lens" is an 8.0 mH inductor that we purchased from Parts Express (catalog number 266–854). Its turns are made of 18 AWG magnet wire that are wound very tightly and bonded together. This is important because it minimizes magnetic distortions that translate into image distortions. The DC resistance of this inductor is 1.6 Ω with a power handling of 300 W.

You should already have some practical experience with magnetic focusing if you measured e/m using Hoag's method with a CRT (Figure 51). As you may remember, the image of a line (formed by scanning electrons back and forth by electrostatic

---

[‡]The best starting point to construct a single-electron interference setup is to acquire a surplus, working TEM. Units are often sold on eBay for $500 to $1,000. If you decide to take on such project, we strongly encourage you to start with a working TEM, because repairing the ultra-high vacuum, power supplies, and electron optics can turn into a difficult and expensive job.

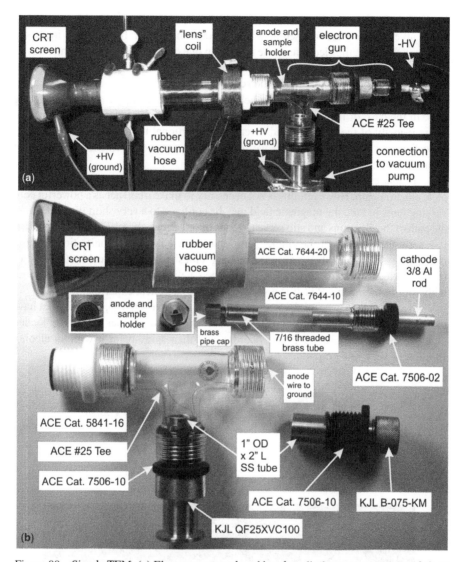

Figure 98   Simple TEM. (a) Electrons are produced by glow discharge at a pressure of about 30 mTorr, directed through a 150-mesh calibration screen, magnified by a magnetic "lens," and projected onto a fluorescent screen. (b) The screen of an oscilloscope CRT is coupled to a 25-mm Ace Glass threaded connector. Ace-Thred glassware and vacuum "Quick-Connec" couplings are used to build the rest of the electron tube.

deflection) is brought to a very sharp focus as the magnetic field along the CRT's axis is increased.

In that experiment, we assumed that all of the electrons would have the same velocity (determined by the accelerating potential applied between the cathode and the anode). We then slowly increased the magnetic field in strength until the time required for the electrons to make one complete revolution due to magnetic deflection is equal to the time for them to travel from the deflecting plates to the screen. At that point, all electrons were focused on a small spot at the screen. If we further increased the magnetic field, the electrons made more than one complete revolution while traveling down the tube. If they made two complete revolutions during this time, they were again brought into sharp focus on the screen.

This behavior is similar to the behavior of light passing through a converging lens. In the case of magnetic focusing, the focal length of the "lens" is proportional to the velocity of the electrons and inversely proportional to the strength of the magnetic field. As such, the solenoid we used in the e/m measurement experiment of Figure 51 is in fact acting as an electron lens.

In that experiment, we wanted a homogeneous magnetic field along the complete electron path. However, this is not necessary to make a coil act as a powerful converging lens. In fact, the magnetic field may be confined to a very short distance along the path of the electrons and still serve to focus the electrons sharply on the screen. This is exactly the purpose of the single "lens" that we use in our demonstration electron microscope. Figure 99 shows our results using a mesh intended for calibration of professional electron microscopes (structure probe catalog number 3011C). Try this out with your homemade CRT! It is a very educational and neat experiment!

If you want to use a biological sample (e.g., a cockroach leg), you will first have to coat it with a thin conductive layer. Attach the sample to a holder with silver-loaded glue (used to repair printed circuit boards), and then spray-paint it with an alcohol solution of very fine graphite.[27]

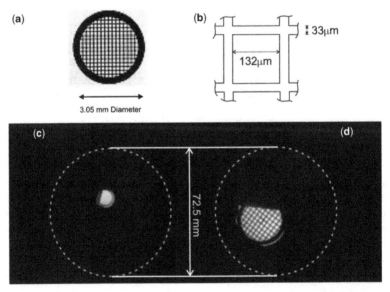

Figure 99   The sample for the simple TEM is an structure probe #3011C mesh used to calibrate professional electron microscopes. (a) Geometry of copper mesh. (b) Detail of grid holes. In the photographs of fluorescent screen shown in (c) and (d), the dashed circle indicates the perimeter of the CRT screen: (c) 3.33 × magnification is achieved with a 1-A coil current, (d) 9 × magnification with a 9-A coil current.

Just as in an optical microscope, higher levels of magnification can be obtained by using a series of lenses instead of a single large magnifying lens. As shown in Figure 100, a lens that acts as an optical "condenser" lens projects high-velocity electrons through the object to be imaged. Some of the electrons are stopped by the object, while others get through—just as with a transilluminated object on an optical microscope's slide. Electrons that pass through the object are focused by the "objective" lens to form an inverted, magnified virtual image. This image is then magnified and reversed again by the "ocular" lens to produce a greatly magnified shadow of the object. In the case of an electron microscope, a fluorescent screen is used to render the electron image visible. Magnifications of 10^5 are common for TEMs using a two-stage system such as this.

More importantly, however, small parts of the object can be distinguished from one another with an electron microscope. The great *resolving power* of the electron microscope is due to the de Broglie wavelength of the electrons, which is smaller than the wavelength of the light in an optical microscope. A two-stage TEM can resolve image features as small as 5 nm, which is more than a hundred times smaller than what can be distinguished with an optical microscope.

Back in 1973, the "Amateur Scientist" column of *Scientific American* reported on the electron microscopes built by a high school physics club.[28] Their simple,

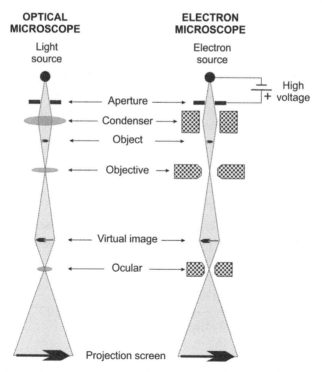

Figure 100 A TEM is very similar to an optical microscope, but it uses electrons and magnetic lenses instead of light and glass lenses. Large amounts of magnification with small image distortion are obtained by amplifying the image in stages, rather than with a single large lens.

one-lens units achieved a magnification of 100×, while their homebuilt, multistage model reached a maximum magnification of 10,000×. Very impressive when you consider that the best optical microscopes top off at around 2,500×!

BLURRING THE LINE BETWEEN QUANTUM AND CLASSICAL

Electrons are still very small particles, so the question arises as to whether quantum weirdness extends to larger objects. The experimental answer is yes!

In 1988, Austrian physicist Anton Zeilinger performed the basic interference experiment with neutrons.[29] In 1991, Carnal and Mlynek did the same with helium atoms.[30] In 1999, Markus Arndt, Anton Zeilinger, and co-workers[31] at the University of Vienna in Austria sent a collimated beam of carbon-60 "buckyball" molecules through a slit made of silicon nitride and detected the interference pattern by ionizing the molecules with a laser and then counting the ions. The slits in the diffraction grating were 50-nm wide and the grating had a period of 100 nm. The central maximum and the two first-order diffraction peaks in the interference pattern were detected with single molecules flying through the apparatus, confirming interference effects predicted by quantum mechanics. The molecules had a most probable velocity of 220 m/s, which corresponds to a de Broglie wavelength of 2.5×10^{-12} m, which is approximately 400 times smaller than the diameter of the C_{60} Buckminsterfullerene molecule.

More recently, the Vienna team performed an improved experiment with fluorinated buckyballs.[32] These biological molecules are present in chlorophyll and have a diameter of about 2 nm, which is over twice as big as a C_{60} Buckminsterfullerene molecule. The team has also reported a successful interference experiment with a fullerene compound that contains 60 carbon and 48 fluorine atoms, making this the most complex object to show two-slit interference to date.

PARTICLE–WAVE DUALITY IN THE MACROSCOPIC WORLD

If all matter has wave-like properties, why is it that we don't observe quantum effects in our daily lives? Consider, for example, the de Broglie wavelength of a 0.15-kg baseball batted at 30 m/s:

$$\lambda = \frac{h}{mv} = \frac{6.626 \times 10^{-34} \left[\frac{\text{kg} \cdot \text{m}^2}{\text{s}}\right]}{0.15[\text{kg}] \times 30\left[\frac{\text{m}}{\text{s}}\right]} = 1.47 \times 10^{-34}[\text{m}]$$

At that wavelength, a baseball would need to interact with objects smaller than subatomic particles to show quantum effects.

In summary, because h is so small, the wavelength of matter is also very small for normal-sized objects with even a tiny bit of momentum. Since the wavelength of these objects is much smaller than the size of objects or systems with which it interacts, we ordinarily don't notice the wave aspect in everyday objects.

However, because wavelength depends on velocity, it can become significant even for macroscopic systems when particle velocities are exceedingly low. This happens at temperatures close to absolute zero (0 Kelvin = −273.15°C), where atoms lose all thermal energy and have only their quantum motion. Massachusetts Institute of Technology researchers recently announced that they were able to cool a dime-sized mirror within one degree of absolute zero.[33] Although this temperature is still too high for observing quantum effects, a technique has been developed to get large objects to ultimately show their quantum behavior. Once the mirror can get cold enough—to within a millidegree Kelvin, quantum effects such "quantum entanglement" between the light and the mirror should be observable. This would confirm that very large objects also obey the laws of quantum mechanics, just as molecules do.

This is not a competition of "mine is larger than yours." The point of these experiments is extremely profound: they demonstrate that quantum behavior is not restricted to the microscopic world. Large, complex macroscopic objects, once thought to be clearly in the domain of classical physics obey the same quantum laws as microscopic particles. This begs the question: is there a limit to the size and complexity of the object that no longer abides by quantum behavior? Probably not!

The outcomes of these experiments have caused an extraordinarily profound impact on physicists' way of thinking. It showed us that, while our classical view of a single reality may work most of the time, the equations of quantum physics describe a tangible, more fundamental level of reality—one that is absolutely incompatible with our day-to-day experience. If classical physics would suffice to describe the behavior of all objects, then we should expect that a single chunk of matter will go through one slit or the other in a two-slit experiment, and will end up in one of two possible bands on a detector. But that is not what happens. Instead, each particle behaves as if it went through both slits simultaneously, not just in theory, but in experimental actuality!

EXPERIMENTS AND QUESTIONS

1. Connect your PMT probe (Figure 29) fitted with a NaI(Tl) scintillation crystal to a high-voltage power supply (Figure 31 or Figure 32) and PMT processor (Figure 34) to a computer running a software MCA. Obtain spectra for exempt, plastic-disk gamma-ray radioactive sources containing ^{137}Cs, ^{60}Co, ^{22}Na, and ^{133}Ba. It is important that the disc source & probe will be placed away from any solid surfaces. Calibrate your MCA using the peaks shown in Figure 87a and Figure 88. Do your measurements show features besides those in the stylized spectra? What may cause other spectral peaks to appear?

2. Set up your ^{137}Cs exempt, plastic-enclosed disk source and your PMT/NaI(Tl) scintillation probe as shown in Figure 87a. It is important that the disk source and probe be placed far from any solid surfaces. Record the gamma spectrum. Next, place a large aluminum block behind the ^{137}Cs source as shown in Figure 87b. Identify the Compton edge and Compton peak produced by Compton Scattering. How well do your measurements compare with the theoretical predictions based on $1/E' - 1/E = (1 - \cos(\theta))/m_e c^2$?

3. Use your measurements of E, E', and E_e using the ^{137}Cs, ^{60}Co, ^{22}Na, and ^{133}Ba sources to estimate the rest mass of the electron. How well does your estimate agree with the accepted value of $m_e = 9.11 \times 10^{-31}$ kg or its equivalent of 511 keV?

4. How does the Compton Effect prove the existence of photons as momentum-carrying particles?

5. What is the momentum of a violet photon at 405 nm? What would happen to the direction and wavelength of this photon after it collides with an electron?

6. Build the single-photon interference apparatus shown in Figure 90 and Figure 93. Calculate the attenuation needed for your laser to ensure that on average only one photon goes through the double slit at any one time. Capture and integrate 1, 5, 10, 25, 50, 100, 250, 500, and 1,000 frames. What is the best evidence your experiment provides that light has a wave-like nature? What is the best evidence your experiment provides that light has a particle-like nature? How are these two pieces of evidence reconciled?

7. Connect a Teltron 555 electron-diffraction tube as shown in Figure 95a and Figure 96a. Measure the diameters of the circles displayed on the tube's screen at accelerating voltages of 2.5–5 kV in 500-V steps. Calculate the de Broglie wavelength for the electrons as a function of their velocity using $D s/2r = \lambda \Rightarrow D = 0.28\,[\text{m}] \times \lambda/s$. How do your measurements compare with the predictions made by de Broglie's hypothesis? How does this experiment demonstrate the wave nature of an electron?

8. Why does the diffracted beam from the target in the Teltron 555 electron-diffraction tube form a circle on the screen rather than distinct points?

9. Why does changing the accelerating voltage in the Teltron 555 electron-diffraction tube affect the diameter of the diffraction rings?

10. The wavelength of a nonrelativistic electron is 0.02 nm. How fast is it moving? Express your answer as a fraction of the speed of light, c.

11. How does de Broglie's hypothesis explain the vastly improved resolution that can be attained by an electron microscope compared to an optical microscope? What is the wavelength of a photon that has an energy of 1eV? What is the de Broglie wavelength for an electron with the same kinetic energy (1eV)? What is the de Broglie wavelength for a 0.15-kg baseball with the same kinetic energy? Why are electrons and photons, but not baseballs, useful particles for exploring the microscopic world?

CHAPTER 6

THE UNCERTAINTY PRINCIPLE

In the last chapter, we presented de Broglie's proposal that wave–particle duality works for matter in the same way it does for light. That is, although light usually behaves as an electromagnetic wave, it sometimes acts as a particle (photon). At the same time, while electrons mostly behave as particles in CRTs, they sometimes exhibit wave-like behavior. We also saw that recent experiments have demonstrated wave–particle duality in much larger and more massive objects than an electron, such as molecules of a fullerene compound incorporating 60 carbon and 48 fluorine atoms.

WAVEFUNCTIONS

Now, this brings up an interesting question: if a sound wave is a vibration of matter, and a photon is a vibration of electric and magnetic fields, what exactly vibrates when matter acts as a wave? The upsetting answer is that there is no directly measurable quantity to correspond to the matter wave itself. This is what physicists call the lack of an *experimental observable.*

That is, while you can use a microphone to measure the vibrations of air pressure that constitute a sound wave, and you can use a radio receiver to measure the electromagnetic oscillations of light waves (remember our microwave interference experiments using a Gunnplexer?), there is no device that can measure a matter wave directly.

Let's think back to the single-electron interference experiments performed by Tonomura (Figure 101). At any one time, we can only detect the fluorescence produced by a single electron hitting the electron microscope's screen. We can't know where the next electron will fall on the screen. We only know that when a large number of electron hits are accumulated, they will build up an interference pattern. That is, for any one electron that hasn't yet reached the screen, all we can tell is the *probability* that the electron will hit a certain part of the screen.

As uncomfortable as this may sound, a matter wave is actually a *probability wave.* It is not directly measurable, and it only allows us to determine where there is a high or a low probability of finding the particle. Probability is proportional to the square of the wave's amplitude, but measuring its square is not the same as measuring the wave itself.*

---

*This is true, if you want to get all philosophical about it, since we obtain the same result by squaring either a positive number or its negative, so there is no direct way to measure the actual sign of a matter wave.

Exploring Quantum Physics Through Hands-On Projects. By David Prutchi and Shanni R. Prutchi
© 2012 John Wiley & Sons, Inc. Published 2012 by John Wiley & Sons, Inc.

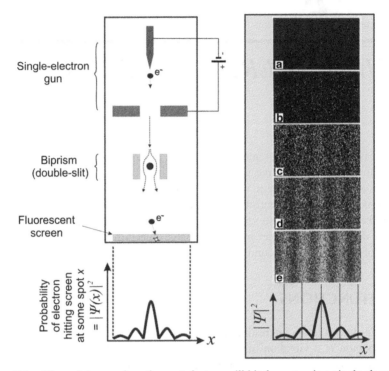

Figure 101   We can't know where the next electron will hit the screen in a single-electron, double-slit experiment. We only know the probability of each electron hitting the screen, such that the interference pattern emerges after accumulating many hits. Inset **(a–e)**: the individual hits in Tonomura's results come at random, showing no apparent pattern at the beginning of the experiment. However, the interference pattern builds up as single electrons arrive more often at locations where the square of the wavefunction peaks than where it dips. Integrated numbers of electrons detected by the detector are **(a)** 10, **(b)** 200, **(c)** 6,000, **(d)** 40,000, and **(e)** 140,000. The inset for this image was obtained from Wikimedia Commons at http://commons. wikimedia.org/wiki/Image:Double-slit_experiment_results_Tanamura_2.jpg. It was released to the public domain with kind permission of Dr. Tonomura. This insert is under the terms of the GNU Free Documentation License.

A matter wave may not be measurable, but as we will see later, quantum physics allows us to calculate its properties for a particle in a physical apparatus. We can then square the mathematical function that defines the matter wave to find the probability of finding the particle at a given place in the apparatus.

We represent the mathematical function for a matter wave by Ψ—the Greek letter psi—and call it the *wavefunction*. Thus, $|\Psi|^2$ is the probability of finding the particle for which we are calculating the wavefunction Ψ at a certain time and position (given by time t and coordinates x, y, and z). For example, in the simple case of the double-slit experiment of Figure 101, $|\Psi|^2$ can be thought of as being dependent only on the position of the screen along the horizontal axis x.

THE UNCERTAINTY PRINCIPLE

In 1925, German physicist Werner Heisenberg was an assistant to Niels Bohr at the Institute of Theoretical Physics at the University of Copenhagen. Heisenberg's research related to the development of mathematical operations to calculate the expected results of experiments on hydrogen atoms based only on observables— that is, using only quantities that could be experimentally measured, either directly or indirectly, such as electron momentum and position and photon wavelength.

Heisenberg noticed that the mathematical operations involving the position and momentum of the electron seemed to indicate that you could not measure both of these at the same time without a compromise in the precision of the measurements. The math required to solve Heisenberg's "Matrix Mechanics" baffled the physicists of the time (we certainly don't intend to put you through it), but there is an easier way of understanding this concept by considering the meaning of various wavefunctions.

Let's take the simplest one—the simple sine wave of wavelength λ associated with an electron with momentum p, for which according to de Broglie, $\lambda = h/p$. This wave extends to infinity just oscillating as a simple sine wave.

What does it mean if an electron has a wavefunction Ψ that is a simple sine that extends to infinity? Well, since $|\Psi|^2$ is the probability of finding the electron at a certain time and position, a wavefunction that is spread throughout space means that the electron could be anywhere. In other words, since the wavefunction has a very precise wavelength λ, we know the electron's momentum very precisely, but as shown in Figure 102a, we don't know where in space it is at all.

The wavefunction that we need to find an electron at a specific position is one that has a high value at one specific location and is zero everywhere else. For example, Figure 102b shows us the probability density of an electron that can most likely be found in the vicinity of x. However, what would be the mathematical description of this wavefunction, and what would it mean for the momentum of the electron?

To answer these questions, we need to know that waves can be combined to form a new wave. Just like the waves that we explored in our ripple tank (chapter 1, Figure 3), we can take two waves and add them up to form a new wave. The addition, or *superposition*, of waves is the essence of the phenomenon of wave interference.

Notice in Figure 103a what happens when we add four waves of different frequency, each of which is uniformly spread out in space. The resulting wave is no longer uniformly spread out, but rather is more concentrated in one place, making it more likely to find the electron at a specific place. As shown in Figure 103b, adding the right component sine waves lets us produce an even more localized wavefunction that approximates the one on Figure 102b.

The localized wavefunction includes many different wavelengths. That means that the electron has an equal number of probable momentums p_1, p_2, p_3, etc., since each simple sine wavefunction of wavelength λ_n that we mixed in has an associated momentum $p_n = h/\lambda_n$.

As such, a wavefunction that localizes the position of a particle very well must allow for a very large number of possible momentums for the particle. So, whenever we know the position of an electron very precisely, we lose precision in our knowledge

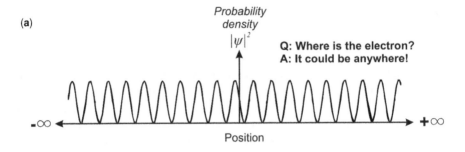

(a)

Probability density $|\psi|^2$

Q: Where is the electron?
A: It could be anywhere!

Position

$-\infty$ $+\infty$

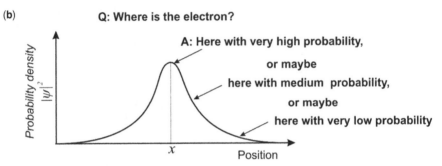

(b)

Q: Where is the electron?

Probability density $|\psi|^2$

A: Here with very high probability,

or maybe

here with medium probability,

or maybe

here with very low probability

x

Position

Figure 102 The probability of finding an electron at a certain position can be calculated from its wavefunction. (a) The electron could be anywhere in space if its wavefunction is a simple sine wave that is spread throughout space. (b) On the other hand, the position of the electron is much more localized for a wavefunction with a probability density $|\psi|^2$ that peaks at a certain position and rapidly decays to zero elsewhere.

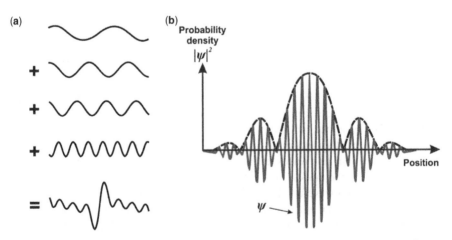

(a)

$+$

$+$

$+$

$=$

(b)

Probability density $|\psi|^2$

Position

$\psi \longrightarrow$

Figure 103 New wavefunctions can be created by adding various simple sine waves of different frequencies and amplitudes. (a) Adding a few sine waves at different frequencies can produce a wave that is more concentrated in a single place. (b) A much more localized wavefunction with a probability density $|\Psi|^2$ (bold dashed line) that peaks at a certain position can be produced by adding many short-wavelength sine wavefunctions together.

of the electron's momentum. This trade-off between simultaneously knowing the position and momentum of a particle is known as the *Uncertainty Principle*. Since quantum physics is the physics of Planck's constant, you may already have a feeling that this constant is at the very heart of the trade-off between simultaneously measuring the position and momentum of a particle, and you would be right! Heisenberg's estimate was that the product of the *uncertainty* in the measurement of position Δx, and the *uncertainty* in the measurement of momentum Δp had to be greater than Planck's constant h. More rigorous analysis resulted in the modern mathematical statement:

$$\Delta x \cdot \Delta p \geq \frac{h}{4\pi}$$

Note that the uncertainties Δx and Δp are not the range of values that x and p can have, but rather the range within one standard deviation of those values.

You will sometimes see the Uncertainty Principle stated as:

$$\Delta x \cdot \Delta p \geq \frac{\hbar}{2}$$

These two forms are identical, because Dirac's constant $\hbar$ (pronounced "h bar," also known as the "reduced Planck constant") is simply $\hbar = h/2\pi$. It is commonly used when frequency is expressed in terms of radians per second ("angular frequency") instead of cycles per second. We don't want to confuse you with this, but rather wish to make it easy for you to understand what the funny $\hbar$ symbol means in other books and papers on quantum physics.

Heisenberg's Uncertainty Principle was published in a 1927 paper that argued why the position and momentum of a particle cannot both be measured exactly, at the same time, even in theory. For Heisenberg, and later for the whole physics society centered on Bohr's Copenhagen school, the very concepts of exact position and exact momentum together, in fact, had no meaning in nature.

Please note that the uncertainty in the measurement of either position or momentum is not due to lack of accuracy of our measurement instruments. The compromise in the measurements happens in principle, as a law of nature. That is, even if your measurement instruments are infinitely precise, you cannot simultaneously know the position and momentum of a particle with absolute precision. For this reason, some people prefer to call this concept the *Indeterminacy Principle*, so that the intrinsic indeterminate character of nature is implied.

Notice, however, that the limitation is in the order of magnitude of Planck's constant h. In fact, since we are dividing it by 4π, it is one order of magnitude lower than Planck's constant. This means that the Uncertainty Principle doesn't really affect our everyday experience, in which momentum is so much larger than Planck's constant.

EXPERIMENTAL DEMONSTRATION OF THE UNCERTAINTY PRINCIPLE

A demonstration of the Uncertainty Principle is simple to conduct using a laser pointer and an adjustable slit consisting of two razor blades. The idea is that the

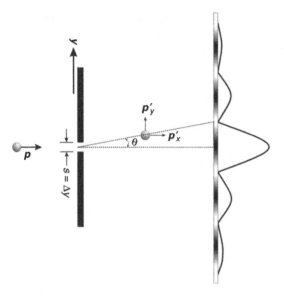

Figure 104 Heisenberg's Uncertainty Principle applied to a particle passing through a slit.

photons emitted by the laser travel without diverging. As shown in Figure 104, the photons exiting the laser can be assumed to have momentum only along the x axis (so $p_y = 0$). However, the exact position of a photon in the beam is unknown.

The position of the photons can be better localized by passing the beam through a slit of width s. This essentially turns the slit into a position-measuring apparatus. However, intercepting the photon beam with a slit will introduce some momentum in the y direction. The more you localize the position of the photon by closing the slit, the more you will become uncertain about the momentum that the photon will have after passing through the slit. This becomes noticeable as a broadening of the diffraction pattern in the y direction, which means that by measuring the position of the photons you have given them some momentum along the y axis that was not there before they encountered the slit.

Given de Broglie's particle–wave relationships, the same exact experiment could be performed with particles that have mass, so let's not restrict the analysis to photons. We'll go slowly so that the math is clear:

For a particle passing through a slit of width s, we know that the particle must be within the slit, so our uncertainty in position along the y axis is:

$$\Delta y = s$$

Before passing through the slit, the particles only have momentum along the x axis (so $p_y = 0$). However, after passing through the slit, the particles will have some momentum component in the y axis ($p_y' \geq 0$).

A diffraction pattern is obtained when we pass light through a single slit. The first minimum in the pattern happens at an angle θ that obeys:

$$\sin(\theta) = \frac{\lambda}{s}$$

so

$$\Delta y = s = \frac{\lambda}{\sin(\theta)}$$

De Broglie tells us the same would happen for any particle, so the particle's uncertainty (one standard deviation) is well approximated by:

$$\Delta p'_y = p \cdot \sin(\theta)$$

The particle's momentum and de Broglie wavelength of a particle are related by:

$$p = \frac{h}{\lambda}$$

so,

$$\Delta p'_y = p \cdot \sin(\theta) = \frac{h}{\lambda}\sin(\theta)$$

Multiply the uncertainty in momentum $\Delta p'_y$ by the uncertainty in position Δy:

$$\Delta y \cdot \Delta p'_y = \left(\frac{\lambda}{\sin(\theta)}\right)\left(\frac{h}{\lambda}\sin(\theta)\right) = h$$

Which was Heisenberg's original estimate for the uncertainty trade-off between position and momentum of a particle!

To conduct this experiment, we made an inexpensive variable slit using a low-cost digital caliper and two single-edge razor blades. We carefully taped each blade to a jaw of the caliper, as shown in Figure 105a, using double-sided tape. We then aimed

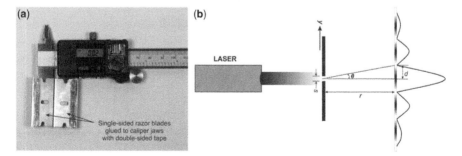

Figure 105　Demonstrating Heisenberg's Uncertainty Principle with photons is easy to do with a laser pointer and a single slit. **(a)** We constructed an inexpensive variable-width slit from a low-cost digital caliper and two single-edge razor blades. **(b)** Measurements of the distance to the first minimum in the diffraction pattern should be taken for a number of slit widths and laser wavelengths.

a laser pointer at the gap between the blades and observed the single-slit diffraction pattern on a distant screen. The pattern scanner that we built in chapter 1 (Figure 5) also works well for measuring single-slit diffraction patterns.

In our experiment, the angle θ is estimated by measuring the position of the first minimum d:

$$\theta = \arctan\left(\frac{d}{r}\right)$$

where d is the measured fringe separation and r is the distance between the slit and the screen.

Which lets us determine the uncertainty in momentum through:

$$\Delta p'_y = \frac{h}{\lambda}\sin(\theta)$$

Since $\Delta y = s$, Heisenberg's Uncertainty Principle is confirmed if:

$$\Delta y \cdot \Delta p'_y = s\frac{h}{\lambda}\sin(\theta) = h$$

or dividing by h, if:

$$\frac{s}{\lambda}\sin(\theta) = 1$$

Go ahead—set up this simple experiment and measure θ for a number of slit widths (e.g., 0.05, 0.1, and 0.2 mm) and laser wavelengths. Using our measurements, we came up with an average value for $s/\lambda \sin(\theta)$ of 1.05, which confirms Heisenberg's Uncertainty Principle within our experimental limits of error.

Another interesting way of conducting this experiment is by using the microwave optics stand we constructed in chapter 1 (Figure 17), placing a single slit in the slide holder, and determining the angles where the minima appear for the system's 2.85-cm-wavelength photons (Figure 106). As the goniometer arms are moved, you should find a first dip in the signal at $\sin(\theta) = \lambda/s$. Good slit widths for this experiment are 7 cm and 13 cm.

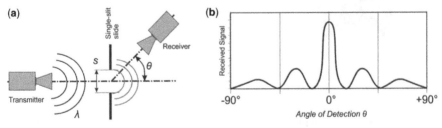

(a) Single-slit slide · Receiver · s · θ · Transmitter · λ

(b) Received Signal · −90° · 0° · +90° · Angle of Detection θ

Figure 106 An interesting way of confirming Heisenberg's Uncertainty Principle is to measure the single-slit diffraction of a microwave beam using the Gunnplexer units we built in chapter 1. **(a)** The angle is measured directly using the goniometer stand. **(b)** With this geometry, the first dip in the diffraction pattern should appear at $s(\sin\theta) = \lambda$.

Experiments using the same basic concept have been used by Zeilinger's team in Austria to prove Heisenberg's Uncertainty Principle with objects as large as fullerene molecules containing 70 carbon atoms.[31]

TIME–ENERGY UNCERTAINTY

Bohr realized that if there is an uncertainty relationship between momentum p and position x:

$$\Delta x \cdot \Delta p \geq \frac{h}{4\pi}$$

which comes from quantizing momentum as a function of wavelength:

$$p = \frac{h}{\lambda}$$

A similar uncertainty principle must then exist between energy E and time t. This is because:

$$E = hf$$

where frequency is the inverse of a time interval $f = 1/t$, so:

$$E = \frac{h}{t}$$

Indeed, an identical relationship exists for the product of the *uncertainty* in the measurement of a particle's energy ΔE, and the *uncertainty* in the measurement of the time Δt at which the particle possesses that energy. The formal derivation of this uncertainty relationship is more complex than deriving the uncertainty between measurements of position and momentum, so we will limit ourselves to stating it as:

$$\Delta t \cdot \Delta E \geq \frac{h}{4\pi}$$

Experimentally proving this form of the Uncertainty Principle is not easy. It requires ultrafast lasers and detectors that are beyond the reach of most casual experimenters' budgets. However, the uncertainty in frequency measurement of a signal as a function of the signal's duration is a convenient and educational model of the quantum time–energy uncertainty relationship.

FOURIER ANALYSIS

When we separated light into its constituent lines, or binned scintillation photons according to their energy, we were performing spectral analysis. Since light and gamma rays most commonly act as waves, what we have been doing is deconstructing a complex electromagnetic wave into many simple sine waves. The same method can

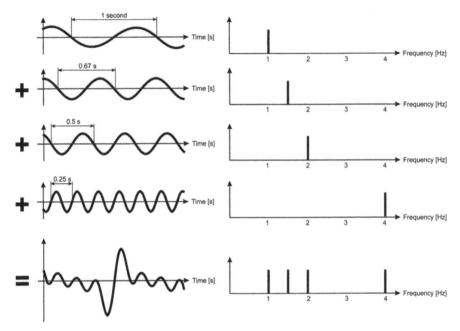

Figure 107   A signal that varies as a function of time (left) can be represented by the frequencies of the sine wave components required to make that signal (right). The spectrum of the signal shows the power of the sine waves that need to be added to produce that signal.

be applied to the analysis of any type of wave, such as sound vibrations, radio signals, or ripples on a pond.

For example, in Figure 103, we saw how we could make a wave with concentrated probability by adding a number of simple sine waves. Figure 107 shows the spectrum of these waves to demonstrate how easy it is to analyze a signal by looking at its sine wave components. This is known as *Fourier analysis* in honor of French mathematician and physicist Joseph Fourier, who discovered in the 1800s that complex waves could be described mathematically as a series of simpler sine waves. The sine wave components of a signal are thus called its *Fourier components*.

Now, to be able to represent any of the waves by a spectrum of neat discrete lines like the ones in Figure 107, we must assume that they repeat in the same exact way over and over forever. However, if we can't analyze the signal for an infinite amount of time (and who has that much time to wait, when there is so much other fun stuff to do?), then we cannot be certain the signal fulfills the requirement of repeating identically forever.

Let's use an audio signal generator to produce an analog of the quantum wave-packet to explore the way in which the Fourier spectrum changes as we vary the length of time available to observe a simple signal. First, download the free ToneBurst[†]

[†]ToneBurst v1.4.0 by David Taylor is available for free download at www.satsignal.eu/software/ audio.html. The required Delphi runtime library can be downloaded from www.satsignal.eu/software/ runtime.html#LibraryBundle.

software by David Taylor from Edinburgh, United Kingdom. You will also need to download the runtime library from his Web site. In addition, download the free Spectrum Lab sound card spectrum analyzer by Wolfgang Buescher.[‡]

Run both programs on the same PC at the same time. Spectrum Lab should display the spectral content of the sound picked up by your PC's microphone. Now, turn up the volume on your speakers, and set ToneBurst to produce a 500-Hz tone with a duration of 500 ms (turn "Declick" on). The spectrum should suddenly show a sharp spike at 500 Hz. Once you verify that this works, experiment with the effect of shortening and lengthening the duration of the tone burst from 50 ms (25 cycles at 500 Hz) up to 1 s (500 cycles at 500 Hz). Do you see how the width of the spectrum sharpens as you increase the burst duration? Play with different settings and try to measure the way in which the width of the spectrum doubles each time you half the duration of the tone burst. (Figure 108)

To understand these results, you must remember that a mathematical sine wave is infinitely long. If you cut off the ends to create a wave packet of finite duration, you have to add many other sine components to cancel the wave beyond the ends of the wave packet. The shorter the segment of the wave you keep, the more additional sine waves you need to cancel the original sine wave outside of the wave packet.

In other words, a sharp function in the time domain yields a very broad function in the frequency domain, and vice versa. This is a form of the Uncertainty Principle and a very important concept in signal analysis—it is not possible to have a very narrow band signal of short duration. In summary, the shorter the wave packet, the broader the spectrum.

The analogy with the time–energy Uncertainty Principle comes from Planck's relationship $E = hf$ (which is valid for both light and particles, as shown by de Broglie). The uncertainty in the measurement of a particle's energy is ΔE, which is analogous in our Fourier spectrum model to Δf. The uncertainty in time Δt is the time during which we can observe the wavepacket.

As we had stated before, the time–energy Uncertainty Principle is expressed as:

$$\Delta t \cdot \Delta E \geq \frac{h}{4\pi}$$

In our model, the wavepacket duration–spectrum uncertainty relationship is:

$$\Delta t \cdot \Delta f = 1$$

We can therefore see that a particle's energy state that exists only for a limited time cannot have a definite energy. To have a definite energy, the de Broglie frequency of the particle must be accurately defined, which requires the state to remain observable for a very large number of cycles.

[‡]Spectrum Lab by Wolfgang Buescher is available for free at www.qsl.net/dl4yhf/spectral.html.

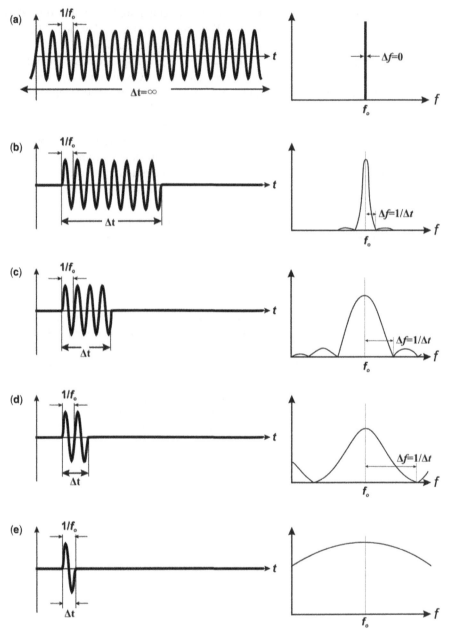

Figure 108   The width of the Fourier spectrum of a truncated sine signal depends on the time available for observing the sine signal. (a) An infinitely long sine signal produces a spectral line of zero width, meaning we have absolute certainty of its frequency. (b) Truncating the wave to an observation time Δt increases the width of the spectrum by $1/\Delta t$ on each side of the sine wave's fundamental frequency f_o. (c–e) The width of the spectrum doubles each time we halve the observation time Δt. Please note that the same scale is not used on every graph for the spectrum power (vertical axis, which does not intercept at $f = 0$).

BYE, BYE CLOCKWORK UNIVERSE

Isaac Newton realized that the laws of motion he had discovered, including the Law of Universal Gravitation, could explain, at least in principle, the behavior of all physical objects in the universe known at the time. These physical laws were so powerful they allowed scientists of the time to calculate with accuracy the movements of bodies in the solar system.

In Newton's mind, the universe resembled a clockwork mechanism. Once built and wound up by God, it wouldn't need God's intervention at all, since it would tick along with its gears governed only by the laws of physics. The *Clockwork Universe Theory* leads to the inescapable philosophical conclusion that in such a mechanistic universe, all things have already been set in motion and are just parts of a predictable machine. Furthermore, since our conscious thoughts are simply the result of physical and chemical interactions between the particles that make the matter in our brains, *free will* is deemed to be an illusion in a clockwork universe. We are simply observers of events that must happen—our choices are simply preprogrammed events meant to happen from the time the clock was wound.

French mathematician and astronomer Pierre-Simon Laplace stated these ideas as the philosophical concept of *Determinism*. In his words:

> We may regard the present state of the universe as the effect of its past and the cause of its future. An intellect which at a certain moment would know all forces that set nature in motion, and all positions of all items of which nature is composed, if this intellect were also vast enough to submit these data to analysis, it would embrace in a single formula the movements of the greatest bodies of the universe and those of the tiniest atom; for such an intellect nothing would be uncertain and the future just like the past would be present before its eyes.

Laplace was imagining what we could now think of as some awesomely powerful computer capable of tracking the position and momentum of every single particle in the universe. Such a computer (known as "Laplace's Demon") would be able to predict the state of the universe at any given time in the future based on simply knowing the position and momentum of all particles at a certain time. Laplace understood that measuring and tracking the position and momentum of every particle in the universe would always be beyond the grasp of human ability, but he saw this merely as a technical limitation, not as a theoretical constraint that would allow humans to have free will.

Let's explore this concept a bit further. Imagine that you have a pool table, and that two balls sit on it. You use a precise laser scanner to measure the position of the balls very exactly. You then hit one of the balls with a pool cue that has been fitted with very accurate force sensors. You also know about the friction between the balls and the pool table's cloth, as well as about the springiness of the table's edges. You would surely agree that a computer fed with all this accurate information can use Newton's laws to calculate the exact position of the balls at any time in the future. Adding one or more balls, or adding forces (e.g., air resistance) complicates the calculations, but does not prevent us—at least in principle—from predicting the future position of all the balls on the table. Divine intervention and consciousness will have no place in these calculations.[§]

Today, space travel owes its tremendous success to the power of these ideas. Indeed, NASA can calculate in advance the exact trajectories its spacecraft must take to rendezvous with distant planets decades after the spacecraft are launched. It is no surprise then that physicists embrace the concept of Determinism and find it very hard to let go of the idea that each particle in the universe must follow a path determined solely by physical laws that govern its motion.

Yet, in 1927, Determinism would be challenged. Heisenberg and Bohr believed that the Uncertainty Principle is a deeply rooted property of nature, limiting in principle what can be known of particles in the universe. As such, Bohr, and his school of quantum physics centered in Copenhagen, believed that even with the most unrealistically accurate measurement instruments it is impossible—even in principle—to determine the state of a system, and that all that one could ever hope to do is to calculate the probability of observing an outcome from an experiment. That is, in a quantum pool table, it is impossible to calculate the exact position of the balls at any time in the future. Even with the most accurate measurement systems and computing power in the universe, all we could ever hope for is to calculate the probability of possible future positions for the quantum balls. By arguing that large objects are made of quantum particles, one can see that the *Copenhagen Interpretation* deals a death blow to Determinism.

On the opposite end, Einstein could not accept the idea that reality at its deepest level would not follow a set of simple physical laws. He believed that the uncertainty posed by Heisenberg is a reflection of our ignorance of some fundamental property of reality, but not a property of reality itself. This is summarized in a famous quote by Einstein that is paraphrased as "God doesn't play dice with the Universe."[¶] Einstein wasn't talking about belief in the will of a Creator,[‖] but rather about nature not following a deterministic path, regardless of its level of complexity.

Einstein was not alone. De Broglie interpreted his matter waves as somehow guiding the motion of particles. This can be compared to the way a surfer rides an ocean wave—we can't see the wave, but the particle indeed feels its effects. This view, espoused by Einstein, de Broglie, and many other scientists of the time is known as the *Pilot Wave Theory*. It was the first type of what is known today as a *Hidden-Variables* theory, which attempts to interpret quantum mechanics as a deterministic theory, avoiding the troublesome paradoxes (such as "Schrödinger's Cat") that we will explore in chapters 7 and 8.

[§]These ideas were very radical in Laplace's time, but the nineteenth century's advanced environment of Enlightenment allowed these thoughts to be explored and discussed openly. For example, in 1802, Laplace gave a copy of his book to Napoleon, who made the embarrassing remark: "M. Laplace, they tell me you have written this large book on the system of the universe, and have never even mentioned its Creator." Laplace answered bluntly "I had no need of that hypothesis."

[¶]Einstein actually said "It seems hard to sneak a look at God's cards. But that He plays dice and uses 'telepathic' methods . . . is something that I cannot believe for a single moment."

[‖]Einstein did not believe in God—at least not by the definition of established religions. In his own words, "I believe in Spinoza's God who reveals himself in the orderly harmony of what exists, not in a God who concerns himself with fates and actions of human beings."

Modern experiments have shown that the universe is indeed a very strange place, and it seems Einstein was wrong about Bohr's interpretation of quantum mechanics. It is very likely that the Uncertainty Principle and wave–particle duality are inherent features of nature, and that there is no way around them. Not for us mortals, and not for any superior intelligence. God does seem to play dice with the universe. Please note that the Copenhagen Interpretation demolishes classical Determinism by introducing true randomness into the machine. However, it does not restore a place for free will, which remains an elusive question best left for philosophers to ponder.

Nevertheless, quantum physics does make some exact predictions. For example, it predicts precisely the energy levels of atomic electrons in absolute agreement with experimental measurements. What quantum mechanics can't predict is the precise path and behavior of individual particles, only the probability that they will be found in a certain place at a certain time.

EXPERIMENTS AND QUESTIONS

1. Build a variable slit using a low-cost digital caliper and two single-edge razor blades, as shown in Figure 105a. Aim laser pointers of different wavelengths (e.g., $\lambda = 630$, 532, and 405 nm) at the gap between the blades, and observe the single-slit diffraction pattern on a distant screen. Calculate $s/\lambda \sin(\theta)$ from measurements using slit widths (s) of 0.05, 0.1, and 0.2mm. What happens as the slit is closed more tightly? Explain your results in the context of quantum mechanics.

2. What would happen if we conducted the same experiment using electrons rather than photons?

3. Place a single-slit slide ($s = 7$ and 13 cm) in the microwave optics stand as shown in Figure 106. Moving the goniometer arms, determine the angles where the minima appear for the system's 2.85-cm-wavelength photons. How well does your measurement of the first dip in the signal match the predicted value of $\sin(\theta) = \lambda/s$ for each slit width? How does this experiment demonstrate the Uncertainty Principle?

4. What does Heisenberg's Uncertainty Principle tell us would happen to Δx for a particle as the particle's velocity approaches zero (i.e., temperature approaches absolute zero)? If the uncertainty in a particle's velocity decreases as it moves slower and slower, does the uncertainty in its position increase, decrease, or stay the same?

5. Run the ToneBurst and Spectrum Lab programs on the same computer, making sure that the spectrum analyzer picks up the signal produced by the tone burst generator. Measure the width of the spectrum produced by a 500-Hz tone burst with durations of 50, 100, 200, 300, 400, 500, and 600 ms, and 1s. What happens to the spectrum each time you double the duration of the tone burst? Why is this a good analogy to the time–energy form of Heisenberg's Uncertainty Principle?

6. The 662-keV gamma-ray line observed from ^{137}Cs (Figure 87) is really produced when its daughter product ^{137m}Ba (excited state of ^{137}Ba) further decays to the ground state of ^{137}Ba. What is the natural spread in gamma ray energy ΔE for this line based on Heisenberg's Uncertainty Principle, if Δt is taken to be the half-life of ^{137m}Ba, which is approximately 153 s?

7. What is the uncertainty in velocity for an electron that has been localized to a position with an uncertainty of 1 nm? what if it were a proton? For your reference, $m_e = 9.11 \times 10^{-31}$ kg and $m_p = 1.67 \times 10^{-27}$ kg.

CHAPTER 7

SCHRÖDINGER (AND HIS ZOMBIE CAT)

In the last chapter we learned that $|\Psi|^2$—the square of a system's wavefunction—is itself a function that gives us the probability of finding a particle at a certain time and position. However, in the last chapter we simply assigned the fancy symbol $|\Psi|^2$ to the probability of finding a particle at a certain place and at a certain time. For example, in the simple case of the double-slit experiment of Figure 101, we just gave the distribution of electron hits on the screen of Tonomura's electron microscope the name $|\Psi|^2$, but we didn't calculate what $|\Psi|^2$ would look like before we looked at the results of the experiment.

In 1925, Austrian physicist Erwin Schrödinger figured out a way of calculating the wavefunction of a system, allowing the behavior of an experimental system to be predicted.

Possibly, the simplest problem that we can solve using Schrödinger's equation is the so-called "particle-in-a-box" problem, in which we determine the probability of finding a particle at a certain position within an ideal box. As shown in Figure 109, a particle of mass m is trapped between two infinitely high walls. Let's assume the box is in the vacuum of space, away from gravitational pull and any other influence. The walls are perfectly elastic, so they don't absorb any energy from the particle. This is modeled by setting the potential energy at zero between the two walls, and making it infinite outside the box:

$$V(x) = \begin{Bmatrix} 0; & 0 \le x \le L \\ \infty; & \text{otherwise} \end{Bmatrix}$$

Mathematically, since the total energy of the particle is given by the sum of its potential and kinetic energies, the particle would need an infinite amount of energy to be outside the box (where $V = \infty$). As such, the particle must be located somewhere between $x = 0$ and $x = L$.

In the classical case of Figure 109a, the particle simply bounces back and forth between the walls at $x = 0$ and $x = L$ with velocity v. Since the box has perfectly reflecting boundaries, every time the particle hits the wall, its velocity is reversed from v to $-v$. If we know the particle's position and momentum at any moment, we can calculate the particle's future position at any time using Newton's equations.

Exploring Quantum Physics Through Hands-On Projects. By David Prutchi and Shanni R. Prutchi
© 2012 John Wiley & Sons, Inc. Published 2012 by John Wiley & Sons, Inc.

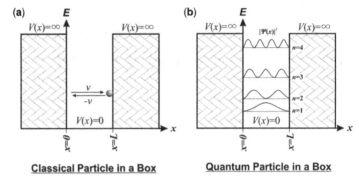

<u>**Classical Particle in a Box**</u>   <u>**Quantum Particle in a Box**</u>

Figure 109   In the particle-in-a-box problem, one attempts to calculate the position of the particle within a box lined by two perfectly elastic, impenetrable walls. **(a)** A classical particle simply bounces back and forth between the walls at a speed related to the particle's energy. **(b)** The position of a quantum particle is unknown until measured, but the probability of finding the particle possessing a certain amount of energy can be calculated with Schrödinger's equation.

The classical momentum and energy of this particle are:

$$p_{\text{classical}} = mv$$

$$E_{\text{classical}} = \frac{1}{2}mv^2 = \frac{p^2}{2m}$$

The particle is allowed to have any nonnegative, noninfinite energy, including zero—in which case it would simply rest at a certain position forever. The important point is that the position and velocity of this classical particle are determined at every instant, regardless of whether it is being observed or not.

For a very small particle however, Heisenberg's Uncertainty Principle will become very significant, and we will not be able to simultaneously know the precise position and precise momentum of the particle at any time. Just like an electron in Bohr's atom, a particle in the one-dimensional box of Figure 109b is in a bound state with a definite energy, but it has no definite position, because of the Uncertainty Principle. As such, the quantum particle within the box has a definite momentum (and hence has definite energy), but its position inside the box is completely random. When it is not observed, the quantum particle exists in a random state, which in quantum physics is called a *virtual state*, whereby the position of the particle—if measured—could be anywhere within the interval $0 \leq x \leq L$. From the previous chapter, we know that if we run the experiment many times we will find a probability distribution that we have called $|\Psi|^2$.

Schrödinger's achievement was to come up with a differential equation to calculate Ψ just from a mathematical definition of the problem. Don't panic if you don't yet know (or no longer remember) calculus! In this book we'll work with the solutions to the equations. In spite of this, let's take a look at Schrödinger's equation for a simple, one-dimensional system that doesn't vary with time. Conveniently, our one-dimensional particle in a box is just this type of simple system. Ready? Here it goes:

$$E\Psi(x) = -\frac{\hbar^2}{2m} \cdot \frac{\partial^2 \Psi(x)}{\partial x^2} + V(x)\Psi(x)$$

Breathe deep. Don't panic. In this equation:

$\Psi(x)$ is the particle's wavefunction as a function of x

E is the energy of the system

m is the mass of the particle

$V(x)$ is the potential experienced by the particle as a function of x, and

$\hbar$ is Dirac's constant, which as you may remember is $\hbar = h/2\pi$.

The $\partial^2 \Psi(x)/\partial x^2$ part is the second derivative of the wavefunction $\Psi(x)$, which is a complicated way of saying that it is a mathematical operation that describes how $\Psi(x)$ changes from one place to another. Don't worry about it, we'll calculate it for you in case your calculus is not up to speed.

Now, we know the following two things that really simplify Schrödinger's equation for our simple particle-in-a-box problem:

1. The particle cannot be outside the box, so $\Psi = 0$ for $x \leq 0$ and $x \geq L$.
2. The potential $V(x)$ experienced by the particle within the box is zero.

So, we can get rid of $V(x)\Psi(x)$, leaving us with:

$$E\Psi(x) = -\frac{\hbar^2}{2m} \cdot \frac{\partial^2 \Psi(x)}{\partial x^2}$$

The solution to this differential equation is:

$$\Psi(x) = \sqrt{\frac{2}{L}} \sin\left(\frac{n\pi x}{L}\right); \text{ where } n = 1, 2, 3, \ldots$$

Notice that the wavefunction is dependent on a sine function, so it will work out for integer multiples of $\pi x/L$, which is why there are multiple wavefunctions for this system that depend on n. As you may realize by now, the reason for multiple solutions at integer values comes from quantization of energy. That is, the trapped particle is allowed to have only discrete levels of energy given by:

$$E_n = \frac{n^2 \pi^2 \hbar^2}{2mL^2}; \text{ where } n = 1, 2, 3, \ldots$$

Notice that n cannot be zero. This is very interesting, because it means that while a classical particle trapped inside a classical box can sit motionless (have zero energy), a quantum particle cannot have zero energy. How is that possible? Well, if the quantum particle would sit motionless, then we would know its exact position. However, according to Heisenberg's Uncertainty Principle, this would mean the particle must have infinite momentum, which is impossible, since the particle is sitting motionless. The opposite idea doesn't work either, since if we say that the particle has zero momentum, we can't know anything at all about its position. However, we know that the particle is somewhere within the finite box, making it impossible for it to have zero momentum . . .

The state of the system for $n = 1$ is known as its *ground state*, and the amount of energy that the system has in this state is known as the *zero-point energy*. This very unintuitive and absolutely fascinating consequence of quantum physics appears with even the simplest quantum system. By the way, you need to be very careful about claims made about devices that supposedly extract useful energy out of the zero-point energy. No device claiming to operate using zero-point energy has ever delivered on its promise.

Let's return to the energy levels that result from Schrödinger's equation:

$$E_n = \frac{n^2 \pi^2 \hbar^2}{2mL^2}; \text{ where } n = 1, 2, 3, \ldots$$

For $n = 1$, the ground state energy is:

$$E_1 = \frac{\pi^2 \hbar^2}{2mL^2}$$

and the energy for all other levels is found by multiplying the zero-point energy by n^2:

$$E_n = n^2 E_1; \text{ where } n = 1, 2, 3, \ldots$$

where n is known as the system's *quantum number*.

Another interesting result from the solution of Schrödinger's equation is that the probability of finding the quantum particle within the box is variable. As shown in Figure 110, classical physics predicts that the probability of finding the particle at any point within the box is equally distributed over the box. Classical probability within the box is simply a line at $1/L$. On the other hand, the quantum probability is not only unevenly distributed, but there are points within the box where the particle

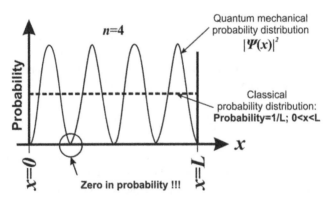

Figure 110   There is a very large difference between the predictions made by classical and quantum physics for the position at which a particle is found in a one-dimensional box. Quantum physics even predicts that there are points at which the particle will never be found inside the box.

could never be found! This might not be too surprising, until you think about how that particle gets from one side of the zero to the other side of the zero. The Copenhagen Interpretation avoids this problem by telling us this is a question that cannot be asked, since "what cannot be observed does not exist."

REAL-WORLD PARTICLE IN A BOX

There are very few real-world systems that behave like the simplified mathematical model we used above for a particle in a box. However, recent developments in the nascent field of *nanotechnology* have made it possible to produce *nanostructures* of only 2–10 nm in size that exhibit behavior close to the simple particle-in-a-box model we discussed.

These tiny structures are known as *quantum dots*. As shown in Figure 111, they are small *semiconductor* crystals of only 10 to 50 times the diameter of an atom. A *semiconductor* is a material in which an electric current can flow due to the movement of electrons and "holes"—states where electrons could sit, but which are empty and thus act as positive charge carriers. You can study about those in most introductory books on electronics. For our purposes, it is sufficient to say that the quantum dots contain free pairs of one electron and one "hole" that can move freely inside the semiconductor, but cannot get out—just like in the quantum particle-in-a-box problem. Quantum dots of different dimensions are currently being manufactured using semiconductor materials such as cadmium selenide/zinc sulfide [CdSe/ZnS] and indium phosphide/zinc sulfide [InP/ZnS].

Do-it-yourself synthesis of quantum dots is possible, and relatively simple procedures using equipment available in undergraduate laboratories have been published,[34,35] but they involve the use of rather toxic materials which we prefer to

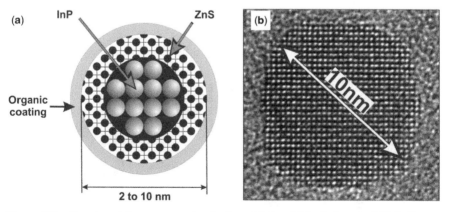

Figure 111 Quantum dots are nanometer-sized crystals in which electrons and "holes" get trapped, acting as real-world models of the particle-in-a-box problem. (a) The quantum dots are made of semiconductor materials such as InP/ZnS, in which the size of the dot can be controlled very well at the time of manufacture. (b) An organized array of atoms is seen in this TEM image of a 10-nm quantum dot.

avoid. Instead, we purchased a set of four vials, each containing a suspension of quantum dots of a specific radius (2.37, 2.53, 2.72, and 2.92 nm). The quantum dots consist of indium phosphide cores surrounded by thin zinc sulfide shells. They are made by Nanosys for Cenco Physics, and are available from educational scientific supplies stores such as Sargent Welch (catalog number WLS1751-18). Another source of ready-made nanocrystal quantum dots is Evident Technologies. They sell kits containing a number of small (although somewhat expensive) vials of CdSe/ZnS and lead sulfide (PbS) quantum dots.

As shown in Figure 112, a quantum particle trapped in a box can absorb a photon (or be excited in some other way), causing the system to move to a higher energy state. However, the higher energy state is unstable, and will eventually decay to the ground state by emitting a photon with energy $E_{photon} = hf = \Delta E$. Similarly, when a quantum dot is illuminated by a photon of sufficiently high energy, an excited electron-hole pair forms inside the quantum dot. When the electron and hole recombine, the quantum dot returns to its ground state, and a photon is emitted with a wavelength dependent on the radius of the quantum dot. Smaller dots emit photons with shorter wavelengths.

Before we experiment with real quantum dots, we must modify the equations to accommodate the fact that the quantum dots are not one-dimensional "boxes," but instead have a spherical geometry. We must also account for the presence of two particles within the box (an electron–hole pair), and consider that the box is not empty, but rather filled with a semiconductor material (which increases the ground state by 2.15×10^{-19} J). We won't go into the derivation of the equation for the InP/ZnS quantum dot, but instead limit ourselves to stating that the energy of the fluorescence (emitted) photon is:

$$E_{photon} = \frac{\pi^2 \hbar^2}{2m_e R^2} + \frac{\pi^2 \hbar^2}{2m_h R^2} + 2.15 \times 10^{-19} \text{ [J]}$$

where

R = radius of the quantum dot

m_e = effective mass of the electron inside the semiconductor = 7.29×10^{-32} kg

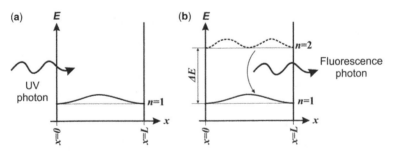

Figure 112   A quantum particle in a box can move to a higher energy state when it absorbs a photon of sufficient energy **(a)**. **(b)** The excited state is not stable, so the system will tend to return to its ground state by emitting a photon with energy equal to the difference in energy between the two states.

m_h = effective mass of the hole inside the semiconductor = 5.47×10^{-31} kg

$$\hbar = \frac{h}{2\pi} = 1.055 \times 10^{-34} \ [\text{J} \cdot \text{s}]$$

To conduct an experiment, illuminate a quantum-dot suspension with a UV LED (at a wavelength of approximately 405 nm) as shown in Figure 113a. Use the spectrometer that you built in chapter 4 (Figure 80) to photograph the spectrum of the light emitted by each quantum-dot suspension. Process the photographed spectra using VisualSpec (as discussed in chapter 4) to obtain spectral plots similar to those of Figure 113b. What is the peak wavelength you measure for each quantum-dot size? How well do your measurements match the theoretical peak emission (Table 6)? Plot the quantum-dot radius versus the peak emission. What type of relationship does this graph show?

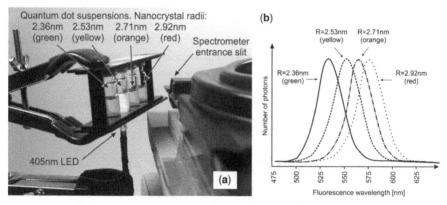

Figure 113   Suspensions of quantum-dot nanocrystals illuminated with UV light behave very similarly to a theoretical quantum particle in a box **(a)**. **(b)** The fluorescence spectra from various quantum-dot suspensions agree with Schrödinger's equation for the quantum particle in a box if adjustments are made to account for the presence of two particles within the box (an electron–hole pair), and for the fact that the "box" is filled with a semiconductor material.

TABLE 6   Nanocrystal Radii and Calculated Fluorescence Wavelength for Quantum-Dot Suspensions[a]

Color	R [nm]	Calculated E_{photon} [J]	Theoretical λ [nm]
Green	2.367454	3.673×10^{-19}	541.1
Yellow	2.533894	3.479×10^{-19}	571.2
Orange	2.718174	3.305×10^{-19}	601.3
Red	2.924941	3.148×10^{-19}	631.4

[a]These data are for quantum-dot suspensions sold by Nanosys/Cenco.

QUANTUM TUNNELING

So far, our analysis of the particle-in-a-box problem has assumed that the barriers around the box are infinite:

$$V(x) = \left\{ \begin{array}{ll} 0; & 0 \leq x \leq L \\ \infty; & \text{otherwise} \end{array} \right\}$$

which allowed us to force the particle to always be within the box. We were thus able to make $\Psi = 0$ for $x \leq 0$ and $x \geq L$.

What would happen if the walls didn't have infinite potential? Well, the interesting thing about waves, which include solutions to Schrödinger's equation for the probability of finding particles at a certain position, is that they are spread-out entities that don't end abruptly when faced with a finite wall or barrier, but instead taper off beyond the obstacle. In the classical mechanics picture of Figure 114a, objects cannot cross regions where their kinetic energy would have to be negative. A ball directed toward a wall will bounce back with no possibility of penetrating the wall. However, as shown in Figure 114b, if the barrier is thin enough, the probability function may extend into the next region. Quantum physics thus predicts there is a chance that a particle trapped behind a barrier without the energy to overcome the barrier may at times appear on the other side of the barrier without overcoming it or breaking it down. This unusual phenomenon is known as "quantum tunneling."

This is completely counterintuitive for macroscopic systems. Imagine that you are trapped in a jail cell (accused of poisoning cats). You don't have enough energy to

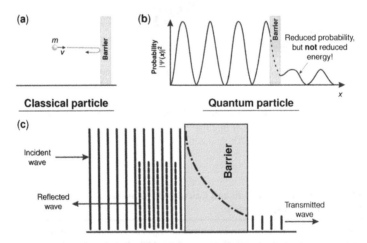

Figure 114   Quantum tunneling through a barrier. (a) In classical physics, a particle with a certain kinetic energy bounces off a barrier that has a higher energy threshold to be overcome. (b) In quantum physics, however, the wavefunction of a particle extends through a barrier, and there is a certain probability of the particle tunneling to the other side of the barrier. (c) For electromagnetic waves (e.g., light), this means that the barrier reflects much of the energy, but allows some of it to tunnel through.

penetrate its walls. Yet, according to quantum physics, there is a certain probability that if you jump against the wall you will disappear from the inside and reappear on the outside leaving the cell's wall unbroken. Quantum tunneling in this macroscopic example is theoretically possible but unimaginably unlikely, yet it occurs all the time for very small particles all around us. For example, alpha particles—which as you may remember from chapter 4 comprise two protons and two neutrons bound together into a particle identical to a helium nucleus—are emitted by an uranium atom when it decays into a thorium atom ($^{238}U \rightarrow ^{234}Th + \alpha$). In classical physics, the alpha particle is forbidden from escaping, because its energy is too weak to overcome the forces trapping it within the nucleus, but according to quantum mechanics, it has a small (but non-zero) probability of "tunneling" through the barrier and appearing on the other side to escape the nucleus. In fact, quantum tunneling is the process behind all radioactive alpha decay. Alpha particles from the ^{241}Am in your smoke detector are constantly being radiated after quantum tunneling out of the nucleus. The tunneling process also happens constantly in the sun: hydrogen nuclei in the sun's core don't have sufficient energy to overcome the "barrier" of electrical repulsion, yet they fuse together to form helium while releasing vast amounts of energy that keep us alive on earth.

As shown in Figure 114c, the same thing happens to an electromagnetic wave hitting a barrier with finite potential, since we may view it as a large group of photons reaching the obstacle. Some of these photons will tunnel through the barrier and reappear behind it.

A simple experiment to demonstrate the bizarre quantum effect of tunneling can be performed with the Gunnplexer microwave units we built in chapter 1 (Figure 12). For this experiment, we must remember that classical optics tells us that electromagnetic waves approaching a dielectric–air boundary at an angle greater than the critical angle for total internal reflection are totally reflected inside the material. For example, if you shine a laser from inside a pool toward the outside of the pool, the laser beam will be completely reflected back into the pool if the laser is aimed beyond a critical angle. Beyond this critical angle there is no refracted wave. This is equivalent to the classical mechanics prohibition of objects crossing regions where their kinetic energy does not suffice to penetrate the barrier (Figure 114a).

When the $\lambda = 2.85$-cm microwaves from a Gunnplexer are directed into a prism made of paraffin or other dielectric material, they will be completely reflected for angles of incidence greater than the critical angle, as shown in Figure 115a. From the classical physics point of view, it is obvious that microwave receiver 2 should not detect any microwave photons from its location behind the reflective barrier. However, a solution to Maxwell's wave equation does exist at the point of microwave receiver 2. The solution indicates that the electric field of the electromagnetic wave will decay exponentially at the interface, rather than oscillate in a sine-wave fashion as it does inside the prism. The meaning of the finite electric field beyond the interface is ignored by classical physics, because it is abstract and seemingly absurd, since Maxwell's equations would show that the electric field doesn't oscillate beyond the reflective barrier.[36]

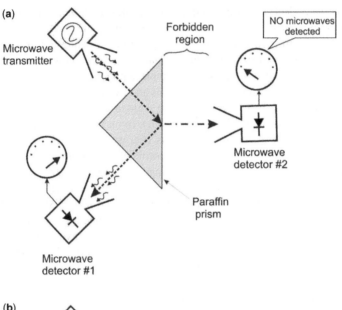

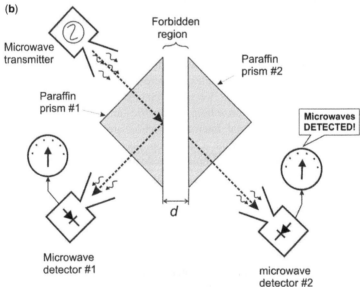

Figure 115 Internal reflection of microwaves (photons with $\lambda \approx 3$ cm) in a prism. **(a)** Normal internal reflection in which photons incident on the paraffin–air interface are completely reflected toward microwave detector 1, allowing no photons to be transmitted for detection by microwave receiver 2. **(b)** Some of the photons tunnel through the forbidden region when a second prism is placed at less than one wavelength away from the reflecting surface.

What is interesting is that placing a second prism in the forbidden region, within one wavelength of the reflective barrier, frustrates the total internal reflection, allowing some of the microwaves to be transmitted through the barrier. From the perspective of

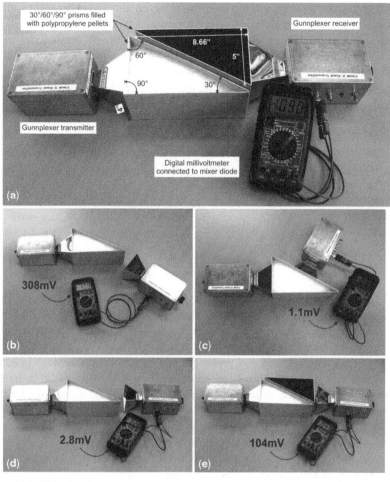

Figure 116   This simple arrangement makes it possible to measure the quantum tunneling of microwave photons through a large air gap. (**a**) We built the frames for our prisms from foam board for the base and top, and thin balsa wood for the walls. The prisms are filled with poly-propylene pellets. The "white" prism has the top removed to show the pellets. (**b**) As expected from a prism, the signal refracts at angle according to Snell's Law. (**c**) and (**d**) Virtually no signal is detected at angles greater than the critical angle. (**e**) However, placing a second prism less than one wavelength away from the reflecting surface frustrates the total internal reflection, allowing some microwave photons to tunnel through the forbidden region.

quantum physics, the wave equation past the barrier is valid and describes the prob-ability of finding photons that tunneled across the barrier. Thus, solving Maxwell's equation inside the second prism produces an oscillating electric field that can be detected by the second microwave detector. This proves that some of the photons must tunnel through the gap between the two prisms (a forbidden region) and reappear when passing through the second prism.

Okay, so let's see how to use the Gunnplexer modules to experiment with photons that vanish and instantly reappear over an inch away! You should build two prisms, as shown in Figure 116a. We first cut a $30°/60°/90°$ foam-board triangle (sides measuring 5-in., 8.66-in., and 10-in.) to use as the base. We made the walls out of $1/16$-in.-thick balsa wood. We then filled each form with around 700 g of polypropylene pellets (roughly $1/8$-in. diameter Poly-Pellets® made by Fairfield, Danbury, CT) sold at the craft store as stuffing for plush toys.

Place one of the prisms on the goniometer stand (the one we built in chapter 1, shown in Figure 17) so the 8.66 in. face rests against the transmitter's horn antenna, and measure the angle at which the goniometer needs to be rotated to detect the strongest signal. This angle should obey Snell's Law, which, as you may remember from the very beginning of the book, states that the ratio of the sines of the angles of incidence and refraction (θ_1, θ_2) is equal to the inverse ratio of the indices of refraction (n_1, n_2):

$$\frac{\sin \theta_1}{\sin \theta_2} = \frac{n_2}{n_1}$$

Assuming that the refraction angle for air is 1.00, what is the index of refraction that you measure for the polypropylene pellets? While scanning with the goniometer, you probably noticed that virtually no signal escapes the prism, except at the prism's specific angle of refraction (Figure 116b).

Place the Gunnplexers and prism on a table, as shown in Figure 116d. You should find that almost no signal can be detected by a receiver placed in line with the transmitter when the prism is in place. In addition, the signal is completely reflected by the prism's long side, as shown in Figure 116c. It is thus impossible to explain how microwave photons should suddenly make an appearance by just adding a prism to the setup of Figure 116d, when the prisms are not even in contact with each other. However, using quantum mechanics, we find that the wave equation past the barrier is valid and describes the probability of finding photons that tunneled across the barrier. Placing a permitted region (the second prism) across the barrier allows photons to tunnel across the forbidden region between the two permitted regions (the two prisms).

QUANTUM TUNNELING TIME

An interesting question to ask is: what is the time that tunneling particles spend inside the barrier? The experimental setup to demonstrate photon tunneling can be modified, as shown in Figure 117, to measure the total tunneling time of the photons, that is $t_v + t_h$. This strange, two-part timing is due to the nature of frustrated total internal reflection. The photon is not a point particle when not being observed, but has a wave equation that extends out into the gap, so the reflection appears to take place behind the surface of the first prism, resulting in a shift down the surface before reflection, D in Figure 117, called the Goos–Hänchen Shift.

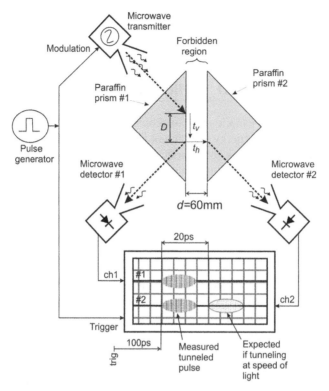

Figure 117   Setup used by Nimtz and Stahlhofen to measure the tunneling time of λ=33 mm photons. This tunneling time has been shown experimentally to exceed the speed of light.

In an experiment carried out by Günter Nimtz and Alfons A. Stahlhofen at the University of Cologne in 2007, a very small value was found for the tunneling time, resulting in velocities faster than light.[37] That is, photons crossing the barrier seem to take less time than they should at light speed. In the actual experiment, using λ = 33-mm microwaves, it was found that both reflected and transmitted beams left their respective prisms at exactly the same time. With the distance d set at 60 mm, the microwaves should have taken 20 ps to cross the gap. However, this time was not detected. The experiment was accurate to ±5 ps, which shows that tunneling inside the barrier happens in zero time! This would mean that a tunneling particle travels faster than light, something that is forbidden by Einstein's Theory of Relativity.

The most common interpretation given by physicists is that only the beginning of the information signal (the leading edge of the "group velocity") moves at speeds faster than the speed of light. However, since the signal itself is smeared through time due to Heisenberg's Uncertainty Principle, a signal that could convey information actually moves no faster than the speed of light, which means that superluminal (faster than the speed of light) communication does not happen, since the actual signal velocity remains less than the speed of light. In spite of this, the experiment by Nimtz and Stahlhofen is claimed by many to be the one and only observed violation of Einstein's Special Relativity. As we will see later, communication at speeds faster than light would enable transmitting information to the past, which opens up a whole set of paradoxes that give physicists a headache.

SUPERPOSITION AND SCHRÖDINGER'S CAT

Let's revisit the original particle-in-a-box problem (Figure 109). We had assumed that the system had a specific energy that did not vary with time. This allowed us to use a simple form of Schrödinger's equation that gave us the wavefunctions of the system that depend only on the position x within the box, and the system's quantum number n:

$$\Psi(x) = \sqrt{\frac{2}{L}}\sin\left(\frac{n\pi x}{L}\right); \text{ where } n = 1, 2, 3, \ldots$$

We then took some quantum dots and illuminated them with UV light (Figure 112) to take them from their ground state ($n = 1$) to their first excited state ($n = 2$). The quantum dots spend some time in the excited state, but eventually return to their ground state by emitting a photon with a wavelength dependent on the radius of the quantum dot.

Since there are a huge number of quantum dots in each vial, some dots are in the ground state ready to absorb UV light, while some are in the excited state ready to decay back to the ground state by emitting a photon. As long as we don't try to observe a quantum dot, all we can know is the probability that it may be in the ground or excited states. However, we saw that each of these states has a different wavefunction, so the wavefunction for the quantum dot exposed to UV light is really shifting between these two wavefunctions. The combined wavefunction is the *superposition* of the two static wavefunctions, and it now varies as a function of time. The time-dependent Schrödinger equation is more difficult to solve, but you should be able to see how it behaves in Figure 118. The probability function is now a blend or superposition of the individual wavefunctions for the system's ground state and first excited state. The probability of finding the particle at a certain position x within the box now also depends on when the box is observed. When the box is not observed, the probability function moves back and forth with time. The physical interpretation of the wavefunction would be that the system is in a super-position of both states until observed. That is, the particle is both at the ground state and in the excited state until observed, at which time the particle may be found at a certain position with probabilities given by the time-dependent Schrödinger equation.

Now, Schrödinger was not the one who interpreted the wavefunction as a wave of probability. He, along with Einstein, absolutely hated the idea that a system would really be in a superposition of states until observed. In opposition to Bohr's group in Copenhagen, Schrödinger and Einstein believed that a particle is in a definite state before we observe it, and the fact that quantum mechanics could only provide an answer in terms of probability meant the theory was incomplete. They were convinced that Determinism would rise triumphant once quantum physics was developed to the

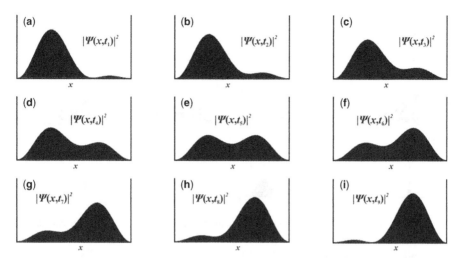

Figure 118 A particle in a box that oscillates between its ground state and first excited state has a wavefunction in which the wavefunctions of each state are superimposed and vary with time. In contrast to a particle in a box at a single quantum state, the probability of finding the particle at a certain position x within the box now also depends on when the box is observed. When the box is not observed, the probability function moves back and forth as time progresses.

point where it could predict the precise state of a system, not only the probability of finding a system in a certain state. In contrast, the Copenhagen Interpretation explains that the system undergoes *collapse* into a definite state only when the system is measured.

Einstein wrote a letter to Schrödinger comparing the superposition of states in a quantum system to the state of an unstable keg of gunpowder that, under the Copenhagen Interpretation, will contain a superposition of both exploded and unexploded states. Schrödinger took this idea to its extreme in his famous thought experiment, attempting to disprove it by stating its absurd consequences. In Schrödinger's words:

> One can even set up quite ridiculous cases. A cat is penned up in a steel chamber, along with the following device (which must be secured against direct interference by the cat): in a Geiger counter, there is a tiny bit of radioactive substance, so small that perhaps in the course of the hour, one of the atoms decays, but also, with equal probability, perhaps none; if it happens, the counter tube discharges, and through a relay releases a hammer that shatters a small flask of hydrocyanic acid. If one has left this entire system to itself for an hour, one would say that the cat still lives if meanwhile no atom has decayed. The wavefunction of the entire system would express this by having in it the living and dead cat (pardon the expression) mixed or smeared out in equal parts.

Since you already know about wavefunctions, we will modify Schrödinger's experiment by tagging the cat's left ear, as shown in Figure 119. This will let us represent the state of the cat by the ear's position along the x axis. If the cat is alive, it will sit and its ear will be at $x = l$. If the cat is dead, then the ear's position will be $x = d$.

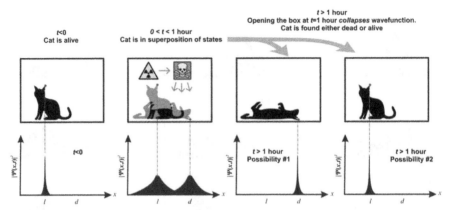

Figure 119   In our version of Schrödinger's cat thought experiment, we tagged the cat's left ear so we can represent the cat's state by the ear's position along the x axis (either at l for live or at d for dead). The cat is definitely alive before we start the experiment at $t = 0$, so the position of the ear is $x = l$. The box is then closed, and Schrödinger's experiment is allowed to proceed, placing both states into superposition. After 1 hour there is exactly 50%/50% probability that the poison did or did not kill the cat, so $|\Psi|^2$ evenly distributes the probability of finding the ear tag at $x = l$ or $x = d$. Once the box is opened, the wavefunction collapses into a certain position at either $x = l$ or $x = d$.

Before this diabolical mechanism is turned on, the cat is definitely alive, so there is absolute certainty that the position of the ear is at $x = l$. The box is then closed, allowing the mechanism to start working. Just as in Schrödinger's original experiment, a radioactive substance inside the box has exactly 50%/50% chance of undergoing one radioactive decay in one hour. Poison is released inside the box if a Geiger counter detects radioactive decay, so the probability of finding the ear tag at $x = l$ decreases as time progresses, while at the same time, the probability of finding it at $x = d$ increases. The wavefunction for the ear's position is thus a superposition of the wavefunction for the cat's state, in which $x = l$ (cat is alive), and the cat's state in which $x = d$ (cat is dead).

The wavefunction would slowly evolve from having a single peak at $x = l$ (cat is definitely alive at $t = 0$), to having peaks of equal height at $x = l$ and $x = d$ (cat is in superposition of states) at the 1-hour mark, when there is exactly 50%/50% chance of the poison having been released. Do you see the similarity between the way in which the cat's ear position wavefunction changes with time and the plots of Figure 118? Figure 118a would correspond to some early time during the experiment, when there is very high probability of finding the cat alive. As time progresses, the probability of finding it alive decreases, while the probability of finding it dead increases, until the wavefunction gets to Figure 118h, in which there is the same probability of finding it dead or alive. Past 1 hour, the probability of finding the cat dead is higher than finding it alive, until the wavefunction of Figure 118l is reached, when there is almost no chance of finding a live cat.

Schrödinger meant this example to be a criticism of the Copenhagen Interpretation of the wavefunction, which implies that the cat remains both alive and dead (to the universe outside the box) until the box is opened. That is, when asked whether the cat is dead or alive during the hour of waiting, the Copenhagen Interpretation would answer that the cat is in a superposition of dead and alive until you look in the box. Then, and only then, does the act of measurement (looking in the box) "collapse the wavefunction," resulting in a cat that's definitely alive or dead.

Einstein and Schrödinger simply could not accept that reality is suspended when it is not being observed. This thought experiment really highlights the main philosophical issues that result from quantum mechanics, including the meaning of the quantum waves, the process of measurement, and the involvement of measuring instruments and observers in processes being studied.

Today, physicists no longer find the process mysterious and intractable, mostly because no physical meaning is given to the wavefunction. Instead, Ψ is understood to be just an abstract mathematical function that contains statistical information about possible experimental outcomes. Whenever the quantum system is measured, the mathematical form of Ψ simply changes, which is common behavior for an abstract mathematical representation, and which doesn't cause any philosophical problems.

MANY-WORLDS INTERPRETATION

The leading alternative interpretation was developed in 1957 by American physicist Hugh Everett, who proposed that the universe splits every time there's an event with more than one possible outcome. Each different universe evolves with one of the possibilities realized. As shown in Figure 120, the bizarre, but logically consistent

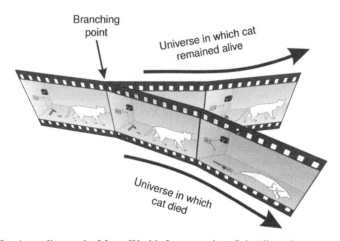

Figure 120   According to the Many-Worlds Interpretation, Schrödinger's cat experiment poses a branching point for the universe. The cat is both alive and dead, irrespective of whether the box is opened, but the "alive" and "dead" cats are in different branches of the universe, both of which are equally real, but which cannot interact with each other. This figure was obtained from WikiMedia Commons, and may be copied, distributed, and/or modified under the terms of the Creative Commons Attribution-ShareAlike 3.0 License.

Many-Worlds Theory interprets the Schrödinger cat thought experiment by explaining that it causes the universe to branch into two separate universes: one in which the cat is alive, and one in which the cat is dead. The cat is both alive and dead, irrespective of whether the box is opened, but the "alive" and "dead" cats are in different branches of the universe, both of which are equally real, but which cannot interact with each other. As such, the Many-Worlds Interpretation requires there to be a very large—perhaps infinite—number of universes, and everything that could possibly have happened in our past, but didn't, has occurred in the past of some other universe or universes.

There are a number of other alternative interpretations of quantum mechanics. Each has its strengths and weaknesses, but they are all outside the scope of this book. In general, we feel that the contrast between Einstein's classical arguments and Bohr's Copenhagen Interpretation provides an excellent framework for our purposes.

SCHRÖDINGER'S CAT IN THE LAB

It is very important to remember that Schrödinger's cat thought experiment was proposed as an absurd extrapolation of the Copenhagen Interpretation. Although we are ignorant about the boundary between quantum and classical systems, quantum physics has very little to do with cats or any other macroscopic system. For all of the reasons that we have discussed so far, real experiments in quantum physics are conducted with very simple systems. In fact, most experiments are conducted with photons using instruments that are just more sensitive versions of the components we have been using in the experiments described in prior chapters.

Real experiments in quantum physics follow the same basic steps as Schrödinger's cat thought experiment. First you need to set up the particle (e.g., cat, photon, electron, etc.) in a superposition of states. This process is called *preparation of quantum state*. The quantum state is then allowed to *evolve*, after which a *measurement* is made to force the system into a certain state (*collapsing* the wavefunction, if you adhere to the Copenhagen Interpretation). The process is commonly repeated over and over again to measure the probabilities of the various outcomes.

The following experiment is not too exciting, but shows all of these basic steps in the simplest possible way. Figure 121a shows the system we used in chapter 2 to detect single photons (Figure 33), to which we have added two pieces of polarizing film. Now, remember that the previous polarization of a photon exiting a polarizer is not important, but its polarization is instead reset to the polarizer's angle (Figure 121b). Therefore, the first polarizer causes all photons entering the system to be polarized at 45°. In quantum physics experiments, this process is thought of as *preparing* the photon's quantum state to a 50%/50% superposition of 0° and 90° polarizations. The photon is then allowed to spend some time in this superposition of quantum states until its quantum state collapses when a *measurement* is made by the analyzing polarizer and PMT. The analyzer is simply a second piece of polarizing film placed within a mount that allows it to rotate (we used a Thorlabs CLR1

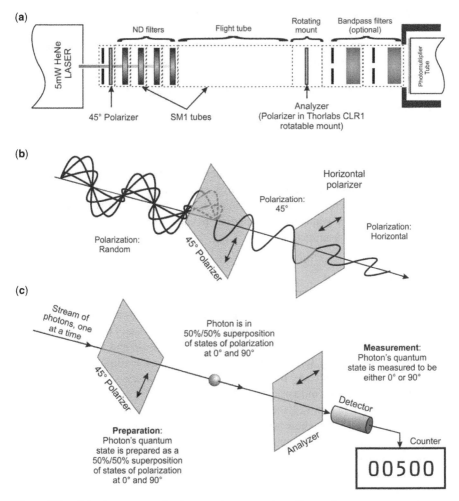

Figure 121   A basic quantum physics experiment consists of preparing a quantum state, allowing the state to evolve, and then conducting a measurement to collapse the system into a state that can be detected. (**a**) In this simple setup, the polarization of photons is set to 45°, which is a 50%/50% superposition of polarization states at 0° and 90°. Photons spend some time as they traverse the flight tube in a superposition of quantum states until their polarization collapses when they reach the measurement instrument comprising the "analyzer" (another piece of polarizing film) and PMT. (**b**) The important thing to remember is that a polarizer doesn't simply allow a photon to pass or not pass, but rather sets the polarization of photons that manage to pass through the polarizer. (**c**) Looking at the process one photon at a time, the first polarizer prepares the photon's quantum state, while the second polarizer (the analyzer), along with the detector, conducts the measurement. The process is commonly repeated many times to determine the probability of occurrence for each possible outcome of the experiment.

SM1-compatible rotatable mount). If a photon exits the analyzing polarizer, it is detected by the PMT (of course, taking into account the absorption of the bandpass filters and quantum efficiency of the PMT).

What is important to understand is that, from the second polarizer's point of view, the photon has two possible polarization states: (1) the photon has the same polarization as the analyzer and can thus go through, or (2) the photon's polarization is orthogonal to that of the polarizer and should be absorbed. The angle between the first and second polarizer only changes the probability of occurrence of each possible outcome. Once detected by the PMT, the collapse finalizes, and we either detect or don't detect the presence of a photon. Do you see how this experiment is analogous to the Schrödinger's Cat *gedankenexperiment* (thought experiment)?

As we will see in the following sections and in chapter 8, the whole field of experimental quantum physics, as well as all of the modern technologies of quantum teleportation, quantum cryptography, and others follow the basic protocol of preparing a quantum state, allowing the state to evolve, and finally performing a measurement. Of course, what differentiates each experiment is how the quantum state is prepared, as well as the things that you can do to a particle in superposition of states without causing collapse of its wavefunction.

BEAM SPLITTERS

Another way in which physicists commonly prepare quantum states is by using partial reflectors as a way of placing a photon in a superposition of states at two different positions. Look out at night through any window in your house and you are essentially looking through a *beam splitter*. This is because you can see light coming through the window from the outside, but you will also be able to see a partial reflection of the light from your own indoor lights.

As shown in Figure 122, the beam splitters we will use are much more reflective than your windows. They allow 50% of the light to be transmitted, and reflect the other 50%. When a single photon encounters the beam splitter, it has a 50% probability of being transmitted and a 50% probability of being reflected.

There are many different materials and optical assemblies that act as beam splitters. Not all of them behave in exactly the same way, so let's take a look at the two best alternatives for our work. As shown in Figure 123, the simplest is a half-silvered or semireflecting mirror. This type of beam splitter, also known as a *plate beam splitter*,

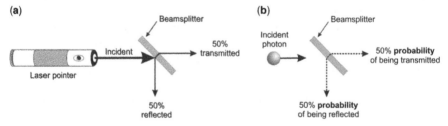

Figure 122   A beam splitter divides a light beam in two. (a) A 50/50 beam splitter equally divides the incoming light intensity into a transmitted beam and a reflected beam. (b) When a single photon encounters the beam splitter, it has a 50% probability of being transmitted and a 50% probability of being reflected.

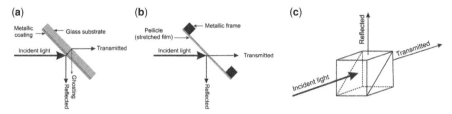

Figure 123   Different materials and optical components can be used to make beam splitters. (a) The simplest is a half-silvered mirror or plate beam splitter, which has the disadvantage of causing ghosting. (b) A pellicle beam splitter is the best choice to reduce ghosting and wavelength sensitivity, but it is very delicate. (c) Cube beam splitters comprise two right-angled prisms with a dielectric coating applied to the hypotenuse surface. Since there is only one reflecting surface in a cube, it inherently avoids ghost images. The nonpolarizing kind is ideal for our experiments at this point.

is simply a plate of glass with a very thin coating of aluminum or another metal of a thickness such that 50% of light incident at a 45° angle is transmitted, and the remainder is reflected. *Pellicle beam splitters* are an improved type of semireflecting mirror that is manufactured by stretching an extremely thin (e.g., 5-μm) polymer membrane over a flat metal frame. The extreme thinness eliminates secondary reflections by making them coincident with the original beam. These membranes are very delicate, but make the beam splitters useful over a very wide range of wavelengths. Stay away from "economy" beam splitters that polarize the output beams in an uncontrolled manner.

The other type of beam splitter useful for our experiments is a glass cube made from two triangular glass prisms glued together at their bases (Figure 123c). In *cube beam splitters*, the glue acts as a dielectric coating capable of reflecting and transmitting a portion of the incident beam. Since there is only one reflecting surface, this design inherently avoids the ghost images that occur with plate-type beam splitters. Please note that cube beam splitters come in polarizing and nonpolarizing versions. You want the nonpolarizing kind for the experiments in this chapter. We'll get to use the polarizing type in the next chapter.

As simple as it is, the beam splitter is a fascinating item to physicists interested in quantum mechanics. For starters, it provides us with the most direct evidence of the particle nature of photons. When we heard the clicks of single photons from a highly-attenuated laser beam striking our PMT probe (Figure 33), we could still claim that light generated by the laser or detected by the PMT comes as tight wavepackets that we call photons. Similarly, in the double-slit experiment with single photons (Figure 90), we could claim that the photon hits detected by the image intensifier were produced by wavepackets that divided at the double-slit slide and then recombined at the intensifier's photocathode. As shown in Figure 124, however, a beam splitter clearly forces a choice on the direction that the photon must take. The resulting detection cannot be the result of recombination of the wavepacket. Only one of the PMTs detects each photon that is sent through the beam splitter. The photon never splits in two to cause detections on both PMTs simultaneously.

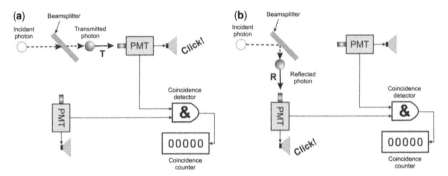

Figure 124 A single-photon beam-splitter experiment provides us with the most direct evidence of the particle nature of light. Photons are either transmitted through the beam splitter **(a)** or reflected by the beam splitter **(b)**, but the photon's wavepacket never splits in two to cause detections on both PMTs simultaneously.

To conduct this experiment, as shown in Figure 125, you will need two PMT probes (Figure 30) and two PMT processing circuits (Figure 34). Neutral-density filters attenuate the laser beam—just as we did before—to a level that ensures only one photon encounters the beam splitter at any single time. The beam splitter randomly allows photons to pass through (which from now on we'll call "T" photons for "transmitted") or reflects them (which from now on we'll call "R" photons for "reflected") with approximately 50%/50% probability. T and R photons are detected separately by

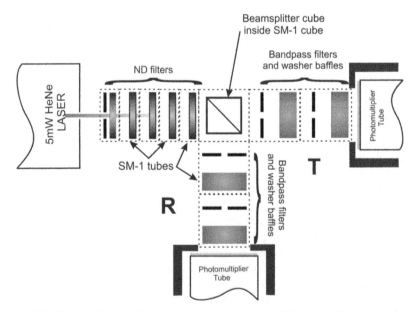

Figure 125 We use this optical setup to demonstrate the particle nature of photons using a beam splitter.

PMT probes. As in the single-photon detection experiment of Figure 33, laser-line filters and washers with $\frac{1}{4}$-in. openings filter out any stray photons that do not come directly from the laser. These are optional if the optical path from the attenuators to the PMTs is completely enclosed within light-tight optical tubes, as shown in Figure 125. However, they are absolutely required if you omit the light-tight tubes and instead build the apparatus inside a darkened box (with the laser outside the box).

We use three Veeder-Root A103-000 totalizer modules to tally the T, R, and coincidence counts. Figure 126 shows the simple AND gate (U1B) we use to detect coincidences. The T and R inputs connect directly to the discriminator outputs of the PMT amplifiers (Figure 34). We adjusted the pulse output width from each of the discriminators (R28) to be around 90 μs. We powered both PMT probes from the same low-ripple, high-voltage power supply.

When you conduct this experiment, adjust the gain of the PMT probes to the same level, and the discriminator thresholds such that clicks are heard only when the laser is on. Reset the counters and run the experiment for a while. You should end up with around 50% of the counts in the T-photon counter, and the other approximately 50% in the R-photon counter. These won't be exact, not only because of

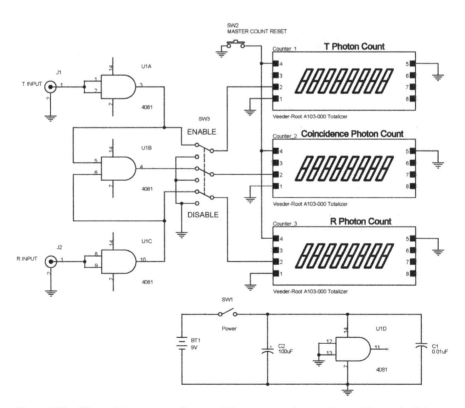

Figure 126   We used low-cost totalizer modules to count photons detected by each of the photomultipliers, as well as to count coincidence events. These modules can count events at a rate of up to 10 kHz. They must be presented with pulses that are least 45-μs wide.

real-world differences in the components used on each leg, but because a random sequence will rarely divide a group exactly down the middle. You should expect the coincidence counts to be low, but you may nevertheless see some counts. These are produced by bunched photons that coincidentally pass through the attenuators at the same time, as well as the coincidental detection of a real photon by one PMT with a "dark count" (noise) by the other.

Coincidence counting is one of the most important techniques used in quantum and particle physics. Coincidence detectors are usually more complicated than the simple AND gate we used, because many experiments require tight control over the "window" during which events must be detected to consider them to be coincidental. In addition, when the window is tightened down to the nanosecond level, controlled delays must be introduced to equalize the timing of the detections. In any case, the analysis of coincidence events always requires a detailed look at statistics. Confidence in coincidence counting is always reduced by uncertainties associated with the statistical timing errors that may occur from the detection process, uncertainties in the electronics, slight changes in gain that skew the time of detection, as well as noise. Nevertheless, this simple coincidence counter should give you a good understanding of the basic concept behind this important technique.

WHO ROLLS THE DICE?

Beam splitters can be classified according to the mechanism used to split the incident light beam. As shown in Figure 127a, some beam splitters are made by depositing reflective elements over a transparent substrate. The reflective elements could be fully reflective mirrors arranged in a polka-dot pattern or fine metallic particles dispersed in a random manner. In these beam splitters, at least in theory, a photon hitting the exact same point on the beam splitter will always do the same thing (either be transmitted or reflected).

On the other hand, cube beam splitters cause photons to be transmitted or reflected completely at random. Even if the same exact spot is hit every time, the photons will randomly transmit or reflect. This begs the question: what is the mechanism

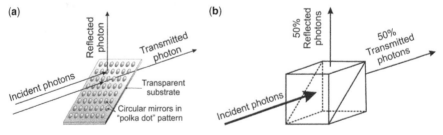

Figure 127   There are two basic techniques used in beam splitters to divide an incident beam. (a) Reflective elements are dispersed over a transparent surface, separating photons based on their location within the incident beam. (b) True quantum beamsplitting through a purely statistical process.

that randomizes the direction that the photons will take when they encounter the beam splitter?

As we discussed before, Einstein, de Broglie, and Schrödinger absolutely abhorred the idea that randomness would still be expected with enough information about the system. So, who is flipping the coin when a photon hits the beam splitter?

One theory that Einstein and de Broglie had proposed to resolve this type of problem is that the photon would somehow be born "preprogrammed" on how to behave when it encounters a beam splitter. The "program" would be completely hidden from our view, so this idea falls within the category of *hidden-variables theory*, about which we will talk widely in chapter 8.

To explore why a simple beam splitter is such an interesting component to physicists, set up three beam splitters as shown in Figure 128. Here, photons are split by a first beam splitter and then again by two other beam splitters, forming four possible paths: TT (transmitted and transmitted again), TR (transmitted and then reflected), RT (reflected and then transmitted), and RR (reflected and reflected again). Now, let's suppose that we follow Einstein's thought and assume that the photons are preprogrammed when they are "born" within the laser to either always transmit or always reflect when encountering a beam splitter. In that case, a T photon coming out of beam splitter 1 would always be transmitted when encountering a beam splitter, so it would exit through TT from beam splitter 2. On the other hand, an R photon coming out of beam splitter 1 would be preprogrammed to always reflect at a beam splitter, so it would have to come out of RR when encountering beam splitter 3. There would never be any photons coming out of TR or RT.

Go ahead—run the experiment by looking at the intensity of the four exit beams. You should find that they are all approximately the same, demonstrating that photons

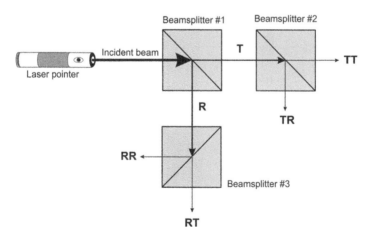

Figure 128   Setting up three beam splitters as shown in this figure defines four possible paths for the photons. T photons exiting beam splitter 1 are split again by beam splitter 2 into TT (transmitted and then transmitted) photons and TR (transmitted and then reflected) photons. R photons exiting beam splitter 1 are also split again by beam splitter 2 into RT and RR photons. If the photons were preprogrammed to always reflect or always transmit when encountering a beam splitter, then we would see photons exiting only at RR and TT but not at RT nor at TR.

are not preprogrammed at birth—at least as far as behavior when encountering a beam splitter. You should find that a photon has a 25% probability of exiting at any given output: 25% for TT, 25% for TR, 25% for RT, and 25% for RR. This simple experiment does not support a hidden-variables theory, and indicates that the "coin flipping" takes place somewhere else within the system to randomize the direction a photon will take when going through a cube beam splitter.

This is a very simple experiment, and you may think it trivial. However, as we will discuss in the next chapter, the question of the existence of hidden variables is at the core of one of the most intriguing properties of quantum mechanics.

THE MACH–ZEHNDER INTERFEROMETER

Let's now complicate the setup just a tiny bit, as shown in Figure 129. Here, photons are split just as before by a first beam splitter into T and R photons. Each of these are then reflected 90° so they will meet at a second beam splitter. Trace the possible paths of the photons and you will see that all TT photons come out of the same exit of beam splitter 2 as all RR photons. The other exit from beam splitter 2 contains all RT and TR photons.

What would be your guess about the probability of a photon coming out of one or the other port of beam splitter 2? Based on the results from the prior experiment (Figure 128), you would probably expect that since RR and TT counts were 25% each, then RR + TT = 50%, and the rest would of course be TR + RT = 50%, right? WRONG!

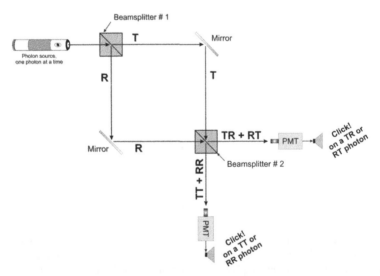

Figure 129   Photons sent through a Mach–Zehnder interferometer are split by a first beam splitter and then recombined by a second beam splitter to yield two outputs: one outputs all TT- and RR-photons, while the second outputs all TR- and RT-photons.

As long as the photon paths between the beam splitters form a perfect square, ALL photons will come out of the TR + RT output, and none will come out of the TT + RR output. This is experimental fact.

Nothing is wrong with the logic. After all, we did analyze how photons interact with beam splitters:

1. The experiment of Figure 125 showed us that photons have approximately 50%/50% probability of being transmitted or reflected by a beam splitter. However, a photon is never split in half. A photon either exits at the T port or the R port of a beam splitter, but never through both at the same time.

2. The experiment of Figure 128 showed us that beam splitters can be stacked, and each simply divides the probability in half. We definitely found 25% of our photons at the TT exit of beam splitter 2, and 25% of the photons at the RR exit of beam splitter 3.

3. Photons in the experiment of Figure 129 were sent one at a time, ensuring that they wouldn't "collide" at the second beam splitter in such a way that they would only exit through one of the ports.

As you may have figured out by now, the reason for the strange behavior is that the Mach–Zehnder interferometer behaves very similarly to a double-slit experiment. Just as in the case of the two-slit experiment with single photons (Figure 90), we are looking at the interference of a photon with itself!

If you now think about light as a wave, you will immediately realize that all that is happening is that destructive interference happens at the RR + TT port, while constructive interference happens at the RT + TR port when the legs of the interferometer are of exactly the same length. Photons would start showing up at the RR + TT port as soon as one of the path lengths changed by a fraction of a wavelength.

This last point is important. The Mach–Zehnder interferometer is sensitive to extremely small variations in path length between the beam splitters. For this reason, building a Mach–Zehnder interferometer requires that precision optical mounts and components be used. In addition, the substrate on which the interferometer is built must be very solid and free of vibration.

A very tight photon beam and small detector apertures are necessary to count photons. However, the Mach–Zehnder interferometer can be made to produce an interference pattern by using a broader beam. In essence, a wide beam spreads the lengths of the interferometer's legs so that the RR + TT and TR + RT conditions are met at different angles at the second beam splitter's ports. This technique is convenient if one prefers to use the presence or absence of an interference pattern as an indicator of collapse of the superposition state instead of counting photons. However, as shown in Figure 130, both methods are equivalent.

As shown in Figure 131, we built our interferometer on a $\frac{1}{2}$-in.-thick aluminum optical breadboard (a surplus Edmund Optics model 56935 18-in. × 18-in. bench plate) that we set on our basement's concrete slab atop two bubble-wrap rolls to isolate the interferometer from environmental vibrations. We used a 5-mW green laser pointer that has a permanent-on mode and good thermal dissipation. Although any visible laser will work, output stability is important to ensure the interference pattern

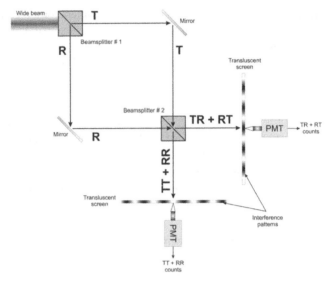

Figure 130   A wide photon beam produces interference patterns at the outputs of beam splitter 2 in a Mach–Zehnder interferometer. If the legs of the interferometer are perfectly balanced, a bright band will be seen directly around the TR + RT output, while a dark band will be observed directly around the TT + RR output. Placing detectors with a very narrow opening directly in front of the beam splitter's outputs produces results identical to those resulting from use of a very narrow beam (Figure 129).

doesn't shift at random. In addition, lasers toward the green portion of the spectrum are best, since they produce easily visible patterns, even at low power. We mounted our laser on an adjustable V-shaped platform mount that gives us independent steering and height adjustments.

In general, the very tight beam produced by a laser pointer is inconveniently small to produce interference fringes that can be seen with ease. For this reason, we expanded the beam to approximately 1-cm in diameter using two lenses, as shown in Figure 131. It is important to adjust the spacing between the lenses to produce a tightly collimated beam that doesn't diverge at a distance of around 10 m. Alternatively, you can use a ready-made 5× or 10× beam expander, but only if you happen to have one on hand, or find it at really low cost in the surplus market.

Our beam splitters are nonpolarizing cubes. We purchased them from Surplus Shed—a great place to find high-quality surplus optical components. Preferably, you should choose beam splitters that are antireflectively coated for the laser wavelength you intend to use. You should also select mounts that allow for adjustment of the position and orientation of the beam splitters (e.g., Thorlabs model KM100B kinematic platform mount with a model PM1 clamping arm).

We use first-surface aluminized mirrors that are flat to one-tenth of a wave over an inch diameter. We found a matching pair of mirrors on eBay that came installed on a pair of Newport MM-1 kinematic mounts. Any similar kinematic mirror mount would work just as well (e.g., Thorlabs model KMM1, Edmund Optics NT58-851, etc.).

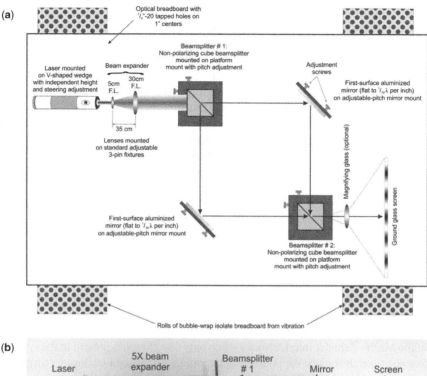

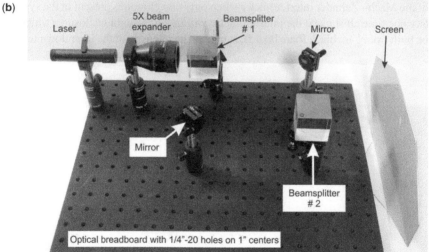

Figure 131 Our Mach–Zehnder interferometer is built on a surplus optical breadboard that sits on our basement's slab atop two rolls of bubble-wrap. (**a**) Good quality optics and solid mountings are necessary to produce a usable interference pattern. (**b**) All of our components are surplus, which explains the unnecessarily large beam-splitter cubes.

Finally, we chose to view the interference fringes produced by the system, instead of counting photons from each port. This makes it similar to the two-slit interference with which we already have some prior experience. To show the interference pattern, we simply project one of the outputs from the second beam splitter onto a piece of ground glass. If necessary, you can use a magnifying glass or any other suitable convex lens to enlarge the interference fringe pattern.

Adjusting the Mach–Zehnder interferometer takes some patience. Start without the magnifying glass or the glass screen, projecting the beam on a surface some 2 m away. You should first adjust one of the mirrors to place its beam right at the center of beam splitter 2. You should then adjust the other mirror so that its beam intersects the first beam at the center of beam splitter 2. Next, adjust the beam splitter until the spots produced by the two beams on the distant surface overlap. Move the surface closer and check again for overlap. Just a few iterations of adjusting the second mirror and beam splitter 2 should suffice. The ground glass screen can then be put in place to visualize the interference pattern.

It is interesting to press very lightly on one of the mirror mounts to change the length of one of the interferometer legs by fractions of a wavelength. This should suffice to disturb the interference pattern without permanently altering the interferometer's setting. You can also try placing a hot soldering iron close to the path of the light on one of the legs of the interferometer. This will change the refractive index of the air, thus changing the speed of light along one of the legs. This should also disturb the interference pattern, demonstrating the extraordinary sensitivity of the interferometer.

Lastly, as shown in Figure 132, it is easy to observe the weird quantum behavior of the Mach–Zehnder interferometer when only one photon is present in the system at any one time. Just place the interferometer within a light-tight enclosure (which may be built from $\frac{1}{4}$-in. black foam board), attenuate the intensity of the laser down to where

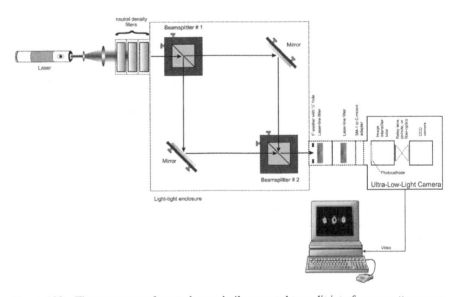

Figure 132   The same type of setup that we built to record two-slit interference patterns one photon at a time (Figure 93) can be used to record the interference pattern produced by flying single photons through the Mach–Zehnder interferometer.

at most one photon is within the interferometer at any one time, and use the makeshift low-light camera and software frame integrator that we built in chapter 5 (Figure 90).

"WHICH-WAY" EXPERIMENTS

Just as with the two-slit experiments, the Mach–Zehnder interferometer builds up an interference pattern even when shooting photons one at a time. Remember that we also learned that Tonomura was able to do the same thing using single electrons, and more recently the team at the University of Vienna demonstrated two-slit interference using a collimated beam of fluorinated carbon-60 "buckyball" molecules.[67] We have repeatedly asked: "Which slit did each of the particles go through?" Invariably, we have reached the conclusion that somehow each particle must behave as if it went through both slits at the same time in order to yield an interference pattern!

Just as in Schrödinger's cat experiment, the particle is forced to be in two mutually exclusive states at the same time. According to the Copenhagen Interpretation, the particle in a two-slit experiment is in a superposition of quantum states until its position is finally measured when it hits the detector. The particle is neither in slit 1 nor in slit 2, but has an equal probability of being found in either one until a measurement is made.

What would happen, then, if we nevertheless try to determine which slit the particle goes through on its way to the screen? An easy way of doing this without affecting the path of the particles would be to tag the particles differently if they go through one or the other slit such that the path of the particle can be identified when it finally reaches the detector.

For example, we could place a vertical polarizer on one slit, and a horizontal polarizer on the other. The surprising result, as shown in Figure 133b, is that if you try to find out through which slit the particle goes, the superposition collapses to a single definite state, and the interference effects vanish.

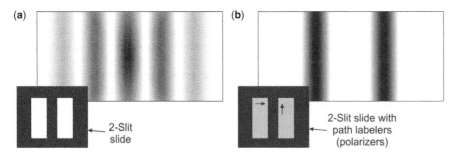

Figure 133   The two-slit experiment can be modified in order to attempt to identify the path taken by photons. (**a**) The unmodified two-slit experiment produces the familiar interference pattern. (**b**) Labeling the paths by tagging the photons with different polarizations depending on which slit they pass through destroys the interference pattern. Note that the width of the slits and interference patterns are largely exaggerated to illustrate the concept.

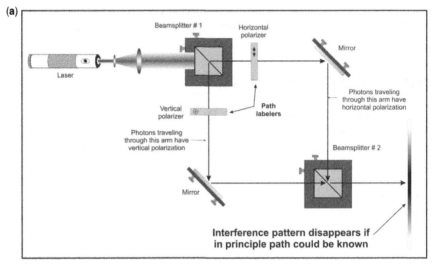

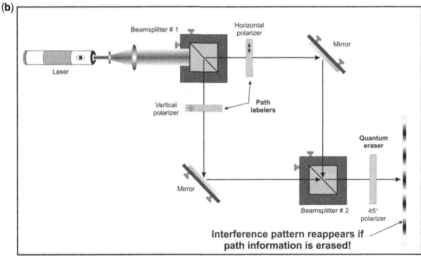

Figure 134 We added polarizers to each of the legs of our Mach–Zehnder interferometer to conduct a which-way experiment. (a) Labeling the paths by tagging the photons with different polarization depending on which leg of the interferometer they take destroys the interference pattern. (b) Insertion of a 45° polarizer in front of the screen erases the which-way information, thus restoring the interference pattern. Note that the width of the interference pattern is largely exaggerated to illustrate the concept.

Placing polarizers on a double-slit slide is not very easy, so let's instead perform this experiment using our Mach–Zehnder interferometer.* Set up your interferometer to produce a clear interference pattern. Then, insert two pieces of linear-polarizing

*If you would nevertheless want to try it with a two-slit setup, you may want to follow the instructions of an excellent article that appeared in the May 2007 issue of *Scientific American.*[38]

film, as shown in Figure 134a. One of the polarizers should be oriented vertically, while the other should be oriented horizontally. This will tag all photons going through one of the paths by forcing them to have vertical polarization, while the photons flying through the other leg of the interferometer are tagged with horizontal polarization. At least in principle, this would allow us to identify the path a photon took through the Mach–Zehnder interferometer before being detected at the screen.

By setting the system in such a way that we may (at least in principle) identify which way the photon traveled, we are essentially performing an experiment in which we observe the particle nature of light. The photons gladly obey, and the interference pattern disappears. We can't simultaneously observe the wave and particle nature of light. The explanation is that the quantum superposition collapses as soon as we measure a property of a photon that is in a superposition of states.

The same exact thing happens when experiments are performed with any other type of particle (electrons, neutrons, atoms, or buckyballs). Measuring the position, velocity, or another property of a particle in a superposition state always gives a definite value. However, at the instant of measurement, the superposition ceases to exist and the particle's wavefunction collapses into a specific state.

For example, Stephan Dürr and his colleagues at the Max Planck Institute for Quantum Optics in Germany conducted a single-atom interferometry experiment in which individual rubidium atoms could be tagged through very subtle, different microwave transitions applied to the two paths.[39] As expected, an interference pattern appeared when no tagging was applied. However, when the tagging equipment was turned on, the interference fringes were lost due to the storage of which-way information in the atoms.

THE QUANTUM ERASER

Let's now see how we would actually go about determining the leg of the interferometer taken by a photon. Let's insert a third polarizer between the output of beam splitter 2 and the ground glass screen, as shown in Figure 134b. This polarizer is the equivalent of the analyzing polarizer of Figure 121. At an angle of $0°$ (horizontal) we are only looking at photons that traveled one leg of the interferometer, and no interference fringes are seen. In the same way, setting the angle of the analyzing polarizer at $90°$ selects photons that have traveled down the other leg, and again no interference pattern is observed. As we have seen before, it isn't even necessary to actually determine the path. The mere existence of which-way information suffices to make the interference fringes disappear.

Now you are in for a real treat. Turn the analyzer to $45°$, as shown in Figure 134b. The interference pattern miraculously reappears! This happens because you have erased the which-way information so you can't—even in principle—determine which path of the interferometer was taken by a photon.[†]

[†]Most books on quantum physics explain the disappearance of the interference pattern in which-way experiments based on a thought experiment proposed by Bohr, in which each slit is illuminated by photons to detect an electron moving through the slit. Bohr then argued that the interference pattern disappears because the particle's direction is modified by the photon that detects it. However, we like the path-labeling method much better, since it encodes which-way information onto the particle without introducing a "kick."

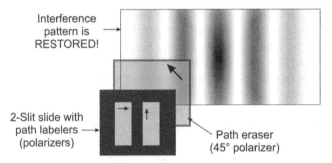

Figure 135   The interference pattern is restored when a quantum eraser is placed before the screen in a traditional two-slit experiment with path labeling (Figure 133b).

As shown in Figure 135, the interference pattern is also restored when a quantum eraser is placed before the screen in a traditional two-slit experiment with path labeling (Figure 133).

It is important to note that the path-labeling process in itself does not disrupt the photons and erase the interference pattern. If the path labeling were responsible for the destruction of the interference pattern, we wouldn't be able to restore it with a quantum eraser. The fact is that the mere existence of the *information* about the path that the photon takes is what destroys the interference pattern.

Another mindboggling result from this experiment is that the interference pattern can be restored by destroying the which-way information after the photon has left the interferometer!

These experiments with quantum erasers produce results that are absolutely baffling. They demonstrate that *information available to the observer* is the critical factor that determines the collapse of a quantum system's wavefunction. In quantum physics, unlike classical physics, the observer and the quantum system being observed become inescapably linked.

A word of caution is in order—the "observer" is not required in any way to be a conscious being. Interaction with a measurement apparatus suffices as an "observation." As Richard Feynman put it: "Nature does not know what you are looking at, and she behaves the way she is going to behave whether you bother to take down the data or not."

EXPERIMENTS AND QUESTIONS

1. Use a 405-nm LED to illuminate suspensions of quantum dots of 2.37, 2.53, 2.72, and 2.92 nm radii, as shown in Figure 113a. Photograph and process the spectrum of the light emitted by each quantum-dot suspension. What is the peak wavelength that you measure for each quantum-dot size? How well do your measurements match the theoretical peak emission based on the solution to the applicable Schrödinger equation for this particle-in-a-box problem?

2. Plot the quantum-dot radius versus the peak emission experiment. What type of relationship does this graph show?

3. Place a 30°/60°/90° prism filled with polypropylene pellets on a goniometer stand such that the 8.66 in. face rests against a Gunnplexer transmitter's horn antenna. Measure the angle at which the strongest signal is detected. Using Snell's Law, and assuming that the refraction angle for air is 1.00, what is the index of refraction that you measure for the polypropylene pellets?

4. Place the prism such that the shortest face (5″) rests against the Gunnplexers transmitter's horn. Plot the signal level detected at all other angles permitted by your goniometer (in 5° increments). How much signal is detected when the two antennas are pointing at each other but the prism is in place? What happens if the prism is removed?

5. Place the Gunnplexers and prism on a table as shown in Figure 116d. Record the signal level, and then place a second prism as shown in Figure 116e. What happens to the detected signal? Explain your results in the context of quantum mechanics.

6. Using the setup of Figure 121a, and using sufficient attenuation to allow at most one photon (in average) to be flying through the optical tube at any one time, measure the number of counts detected by the PMT as the analyzing polarizer is rotated from 0° through 180° (in 5° steps). Explain your results in the context of quantum mechanics.

7. Construct the setup of Figure 125 and feed the outputs of the two PMT signal processors to individual event counters and a coincidence counter. How do the photon-count statistics demonstrate the particle nature of light? Why is this experiment better for this purpose than using a single PMT?

8. Research the way in which a positron emission tomography (PET) scanner works. How is the coincidence-detection technique used in the PET scanner to provide high-resolution images of body organs?

9. Illuminate three beam splitters as shown in Figure 128. Observe the intensity of the light exiting the TT, RR, TR, and RT ports. How do these compare to each other? What would you expect the intensities to be if the individual photons were "preprogrammed" to always behave in the same way (either transmit or reflect) when encountering a beam splitter? What do your experimental results tell you about the behavior of photons when encountering a cube-type beam splitter?

10. Construct and align the Mach–Zehnder interferometer shown in Figure 131. Why is the appearance of interference fringes surprising when this setup is analyzed from the perspective of light as a stream of particles?

11. Push one of the mirror posts very lightly. How does the interference pattern change? Why is this behavior obvious when assuming the wave nature of light?

12. "Label" the paths in the Mach–Zehnder interferometer using polarizing film, as shown in Figure 134a. What happens to the interference pattern? Explain your results in the context of quantum mechanics.

13. Turn one of the polarizers by 90° so both polarizers are oriented in the same direction. What happens? Explain your results in the context of quantum mechanics.

14. Return the polarizers so each path is clearly labeled (polarizers are in orthogonal directions), causing the interference pattern to disappear. Place a third polarizer at the output of the beam splitter, as shown in Figure 134b. What happens? Why is this result surprising? How does this experiment support the Copenhagen Interpretation?

CHAPTER 8

ENTANGLEMENT

We ended the last chapter with the quantum eraser experiment, which seems to indicate that the photon "knows" when we—the observers*—are watching. Indeed, the photon behaves very differently when we—the observers—can say (at least in principle) which path it has traveled.

The thought that an objective reality does not exist independently of an observer troubled Einstein very much. In opposition to Bohr's group in Copenhagen, Einstein believed that the fact that quantum mechanics could only provide an answer in terms of probability meant the theory was incomplete. This was a lively discussion he maintained with Bohr for many years.[†]

In 1935, Einstein refined the philosophical discussion into a physical argument. At Princeton University, Einstein and his assistants Boris Podolsky and Nathan Rosen authored a paper titled "Can Quantum-Mechanical Description of Physical Reality Be Considered Complete?"[40] This now-famous paper, better known simply as "the EPR paper" (for Einstein–Podolsky–Rosen), proposed a thought experiment they felt revealed an underlying, objective reality independent of measurement. The EPR argument seemed to completely contradict the Copenhagen Interpretation.

At the heart of the EPR thought experiment is a particle source that produces pairs of particles that have some property that is forever linked. For example, Figure 136a shows what happens when a subatomic particle called a *pi meson* (also known as a *pion*) decays into an electron and its antiparticle—a positron.

Now, along with mass and charge, electrons have a property known as *spin*. Although electrons don't really spin like tops, the analogy is appropriate, because an electron's spin is related to the electron's angular momentum. Like energy (or its mass equivalent), charge, and momentum, spin is also a conserved amount. Since the spin of the pion is zero, the sum of the spins of the electron and positron produced during the pion's decay must also be zero. Thus, the electron and positron must travel in exactly opposite directions and must have exactly opposite spins.

---

*Please remember that a measurement instrument suffices as an "observer." A conscious being observing the instrument is not necessary for the observation to be completed.

[†]We must mention however, that despite their differences of opinion regarding quantum mechanics, Bohr and Einstein had a mutual admiration that was to last the rest of their lives.

Exploring Quantum Physics Through Hands-On Projects. By David Prutchi and Shanni R. Prutchi
© 2012 John Wiley & Sons, Inc. Published 2012 by John Wiley & Sons, Inc.

(a)

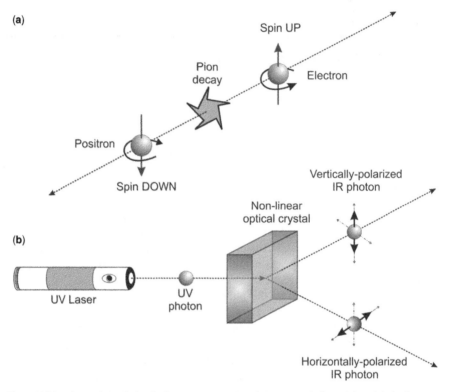

Figure 136   A number of physical processes can produce entangled particles. **(a)** A pion can decay into an electron–positron pair that travel in opposite directions and have opposite spin. **(b)** A UV photon can be converted by a nonlinear optical crystal into two IR photons with opposite polarization.

In a similar way, and as shown in Figure 136b, a UV photon can be converted by a type of crystal known as a nonlinear optical crystal into two IR photons of opposite polarization. The energy is conserved, because each of the IR photons has one-half the energy of the UV photon. Although each pair will come out of the crystal with a random polarization, the polarizations of the photons within the pair will forever be linked due to conservation of angular momentum.

The particles generated in pairs by these processes are said to be *entangled*. According to the Copenhagen Interpretation, two entangled particles share a single wavefunction that is not separable into distinct wavefunctions for each particle. That is, the two particles are linked in such a way that the quantum state of one particle cannot be adequately described without full mention of the other particle, even if the individual particles are separated by a great distance.

Let's suppose that you have a pair of entangled particles (for example, an electron–positron pair) separated by a significant distance. Now, say you measure the electron's spin. This automatically tells you the positron's spin. Since you have managed to ascertain the positron's spin without disturbing the positron in any way, Einstein,

Podolsky, and Rosen argued that it's impossible that the positron only came to have the known spin when you measured it, because you didn't measure it! EPR thus argued that the positron must have had that spin all along.

That is the key argument behind the EPR paper: that measurements conducted on one of the particles of an entangled pair immediately yields information about the state of the other particle, even if they are separated by a significant distance in space. EPR then claimed that the fact that the state of the positron could be determined without interacting with it meant that its spin had an objective reality regardless of measurement.

In Einstein's view, the spins of the two particles are determined together, at the moment they are created. Otherwise, the result of the measurement on the electron would have to be communicated instantaneously to the positron—something Einstein called *"spooky action at a distance."* This was especially disturbing to Einstein, since the instantaneous transfer of information would break the rule of signaling faster than the speed of light, which would violate Special Relativity.

Even more problematic for the Copenhagen Interpretation, a pair of entangled particles as described by Einstein could be used to overcome Heisenberg's Uncertainty Principle. Remember that, according to this principle, you can't simultaneously measure the position and momentum of a particle with absolute certainty. However, let's suppose that you prepare a pair of particles with entangled momentum. Measuring the position of one particle would allow us to know the exact position of the other particle (since they were created together and travel in exactly opposite directions), while measuring the second particle's momentum would allow us to know the first particle's momentum. If the entangled particles are not in communication via some "spooky telepathy," then we would indeed be able to simultaneously know a particle's exact position and momentum.

You may have figured out by now that if the EPR argument is correct, using the entangled-photon source of Figure 136b in a double Mach–Zehnder interferometer would allow us to observe an interference pattern (wave-like behavior) and at the same time determine the path taken by a photon (particle-like behavior), in violation of wave–particle duality.[41]

BELL'S INEQUALITIES

The debate between Einstein and Bohr continued for decades regarding what "reality" meant in the context of quantum mechanics. Einstein and his followers insisted that an objective reality exists whether it is observed or not. Their most powerful argument was explained in the EPR paper, in which Einstein and his colleagues proposed that "elements of reality" (hidden variables) must be added to quantum mechanics to explain the behavior of entangled particles, negating the concept of quantum entanglement, which seems to require action at a distance. In EPR's view, the entangled particles must be born with certain preprogrammed behaviors (those hidden variables) that make them act appropriately when measured without having to communicate with their twins.

These hidden variables would be microscopic properties of particles we are unable to observe directly by means of experimental measurement. Einstein anticipated that we might have the technology to measure them in the future, but for the time being they are "hidden." EPR assumed that if we knew more about these hidden variables, we could finally explain the otherwise mysterious behavior of particles demonstrated by quantum mechanics.

On the other hand, Bohr and his followers insisted that the Uncertainty Principle would remain valid for an entangled system, and that the entangled particles would somehow have to interact at the moment one of a pair is measured, even if they are separated by a vast amount of space. True to the Copenhagen Interpretation, Bohr insisted that these variables are not just unobservable, but rather that they simply don't exist outside of the context of an observation.

This discussion was believed to be purely philosophical and irresolvable, since an experiment would require measurements to be made, yielding the same exact results when an observer is introduced into the picture. In Einstein's words:

> I think that a particle must have a separate reality independent of the measurements. That is: an electron has spin, location and so forth even when it is not being measured. I like to think that the Moon is there even if I am not looking at it.

So, is the Moon there when no one is looking at it? Obviously, this is a purely philosophical question that seems to be impossible to answer.

In 1964, nine years after Einstein's death, Irish physicist John Bell figured out that each of the alternative answers to the EPR paradox would actually show subtle differences in some particular experiments. Specifically, Bell found that if local hidden variables exist,[‡] then a minimum level of coincident detections for the entangled particles would be obtained when certain specific measurements were made. If, on the other hand, quantum entanglement is correct, the number of coincident counts would be below this same level.

The concept is difficult to explain without heavy math. However, Cornell University's N. David Mermin has written a number of ingenious papers in which he explains Bell's Theorem in an easy-to-understand way. We will make use of Mermin's simplified explanations, but we would like to strongly encourage you to read Mermin's original papers to understand the various subtleties of Bell's argument.[42-44] Before we start, remember that you can only measure the polarization of a photon once. This is because, the photon will acquire the polarizer's polarization as soon it is measured (Figure 121b). The "measurement" of a photon's polarization only gives us a "yes" or "no" type of answer based on the probability of the photon passing through or being absorbed by the polarizer according to its angle of polarization. The actual angle of polarization can only be measured to a certain value for a large number of photons.

[‡]Bell's argument relates to *local* hidden variables. That is, hidden variables that are "carried" by each particle or present at the site where particle interactions and measurements take place. This differentiates them from *nonlocal* hidden variables, which are the basis for an alternative interpretation of quantum mechanics of the type outlined by David Bohm. We will not discuss these in this book.

Okay. Following Bell's argument,[§] let's pose the following questions about a single photon:

1. Does the photon have a definite polarization at 0°?
2. Does the photon have a definite polarization at 120°?
3. Does the photon have a definite polarization at 240°?

According to EPR, the photon's polarizations at these three angles are linked to some element of reality (a hidden variable), if they can be predicted without disturbing the photon in any way. We could ascertain the photon's polarization at any one angle, so we could determine its element of reality for that one angle. However, the million-dollar question is whether the photon has elements of reality *simultaneously* at 0°, 120°, and 240°. As we know, we can only measure the photon's polarization once, so we can't know what the photon would have done when encountering polarizers at the other two angles.

According to EPR, elements of reality exist to define the behavior of the photon when encountering a polarizer at 0°, 120°, and 240°, even though we can't actually measure more than one of them at a time. So, at an angle 0°, there may be a hidden variable A; at 120°, a hidden variable B; and at 240°, a hidden variable C. Einstein's followers imagined that one day we may be able to measure these hidden variables and be able to predict with exactitude what a photon would do when it meets a polarizer at 0°, 120°, and 240°.

Figure 137 shows the thought experiment proposed by John Bell as a way of capturing EPR's view on reality. Let's suppose that a photon has three hidden variables A, B, and C. Variable A defines how the photon will behave when it encounters a polarizer at 0°, same for B for a polarizer at 120°, and C for a polarizer at 240°. That is, hidden variable A tells the photon whether or not to cross through a polarizer set at 0°, and so on for B and C. Somehow, these variables would be preprogrammed into the photon at the time it is created, and do not depend at the time of their "programming" on the condition of the polarizer the photon will encounter in the future.

Bell has then defined reality by simply telling us that the photon's polarizations at 0°, 120°, and 240° exist whether or not we try to measure them. Table 7 shows the eight possible permutations of A, B, and C. A photon would thus have to be "programmed" with one of these.

Let's now replace the single-photon generator of Figure 137 by an entangled-photon generator, as shown in Figure 138. Let's also assume the machine produces pairs of photons at a regular interval and detectors placed behind the polarizers report on detections when photons would be expected. This way, not detecting a photon at its expected time of arrival actually means it was absorbed by the polarizer, and not that the photon didn't exist.

According to Bell's interpretation of the EPR argument, both photons would be programmed with the same hidden variables, so this setup allows us to conduct an

[§]Bell originally developed his theorem for two minus one half spin particles, such as protons, neutrons, and electrons. Strictly speaking, the argument is presented for the direction of spin in such particles, and not for photons. However, to keep it simple, we'll assume—almost correctly—that the argument without modification is also valid for the photon's polarization.

(a)

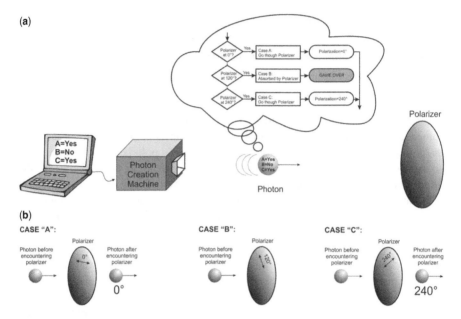

(b)

CASE "A":

Polarizer

Photon before encountering polarizer → $0°$ → Photon after encountering polarizer

$0°$

CASE "B":

Polarizer

Photon before encountering polarizer → $120°$ →

CASE "C":

Polarizer

Photon before encountering polarizer → $240°$ → Photon after encountering polarizer

$240°$

Figure 137   A simplified form of John Bell's view of EPR's argument assumes that a photon has three hidden variables A, B, and C. Variable A defines how the photon will behave when it encounters a polarizer at $0°$, same for B for a polarizer at $120°$, and C for a polarizer at $240°$. **(a)** Somehow, these variables would be preprogrammed into the photon at the time it is created, and do not depend at the time of their "programming" on the condition of the polarizer the photon will encounter in the future. **(b)** We can only measure the photon's polarization once, so we can't know what the photon would have done when encountering polarizers at the other two angles. However, EPR would have asserted that the photon has elements of reality simultaneously at $0°$, $120°$, and $240°$.

TABLE 7   The Eight Possible "Programs" that a Photon may have as Hidden Variables in Bell's Definition of EPR's Reality[a]

Case	A (Does photon pass through polarizer at $0°$?)	B (Does photon pass through polarizer at $120°$?)	C (Does photon pass through polarizer at $240°$?)
1	YES	YES	YES
2	YES	YES	NO
3	YES	NO	YES
4	YES	NO	NO
5	NO	YES	YES
6	NO	YES	NO
7	NO	NO	YES
8	NO	NO	NO

[a]See also Figure 137.

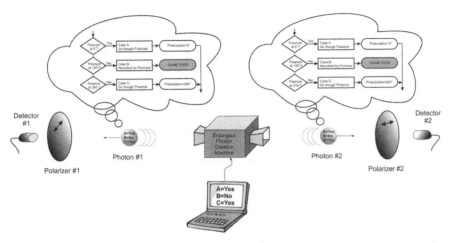

Figure 138   Using an entangled-photon generator instead of the single-photon generator of Figure 137 allows us to test for two of the three hidden variables simultaneously, since, according to EPR, both photons would carry the same "program."

experiment equivalent to making two measurements on the same photon. In this way, we can at least perform a direct measurement on two out of the three hidden variables.

Obviously, the behavior for cases A, B, and C (Table 7) are the same for both polarizers. However, let's rework our table to show coincident measurements for the various possible cases of hidden variables and polarizer settings. Let's use binary notation for the outcomes (YES = 1, NO = 0), and please note that two zeros (neither of the photons went through its polarizer or, in other words, both photons were absorbed by their respective polarizers) also constitutes a coincidence (see Table 8).

Now, please pay attention, because this is the really important part: Table 8 shows us that regardless how the photons are programmed, *the average probability*

TABLE 8   Probability of Coincident Measurements on a Pair of Entangled Photons by Two Independent Analyzing Polarizers[a]

Case	A	B	C	[AB]	[BC]	[AC]	[AB] + [BC] + [AC]	Average Probability
1	1	1	1	1	1	1	3	1
2	1	1	0	1	0	0	1	.333
3	1	0	1	0	0	1	1	.333
4	1	0	0	0	1*	0	1	.333
5	0	1	1	0	1	0	1	.333
6	0	1	0	0	0	1*	1	.333
7	0	0	1	1*	0	0	1	.333
8	0	0	0	1*	1*	1*	3	1

[a][AB], [BC], and [AC] denote coincidences. Asterisks (*) denote a coincident lack of detection (that is, both photons were absorbed by their respective polarizers). If the photon has hidden variables, then the minimum average probability of coincident measurements is at least 0.3333, regardless of the combination of hidden variables and polarizer settings.

of detecting a coincidence is at least 0.333. That is, regardless of how we set up our experimental apparatus of Figure 138, if we perform many measurements, we should end up detecting coincidences between detectors placed behind the polarizers at least one-third of the time. The coincidence detection rate should never be less than one-third as long as you check cases [AB], [BC], and [AC] evenly. The rate should actually be a bit higher than one-third because of the existence of cases 1 and 8.

Now, let's suppose Bohr was right, and hidden variables don't exist. As we will see, Bell chose polarizer angles 120° apart because quantum mechanics (without hidden variables) would predict[†] that the coincidence detection rate for either [AB], [BC], or [AC] is equal to the cosine square of the angle between the two polarizers. Since $\cos^2(120°) = \cos^2(240°) = 0.25$, quantum mechanics would predict that the average probability of coincident detections is around 0.25.

Bell's Theorem, then, tells us that the probability of coincident detections for any local hidden-variables theory (probability > 0.333) is incompatible with the predictions of quantum mechanics (probability ≈ 0.25). Bell had found a way of telling whether the moon is there when nobody is looking! Well ... maybe that's too grandiose. However, Bell's Theorem provides a way of knowing whether Einstein was right or wrong about the existence of local hidden variables.

Bell's test for the existence of hidden variables is thus simply checking that a large number of measurements in an experiment conducted as we have described above will yield a probability of coincident detections above one-third. This is known as *Bell's Inequality*. On the other hand, experimental results violating Bell's Inequality with a sufficiently large statistical confidence would indicate hidden variables don't exist, giving weight to the Copenhagen Interpretation.

In 1969, American physicists John Clauser, Michael Horne, Abner Shimony, and Richard Holt proposed a different version of Bell's Inequality (now known as the *CHSH Test* from the initials of the authors' last names) that is more suited for performing actual experiments to distinguish between the entanglement hypothesis of quantum mechanics and local hidden-variable theories. Since then, other Bell-like inequalities have been proposed to close experimental shortcomings and loopholes. However, as for the original test proposed by Bell, these refined tests are all based on the statistics of counting coincidences between detections. In some of these inequalities, the coincidence limit is a lower bound, like in Bell's original inequality, but others define it as an upper bound, indicating that quantum entanglement would give particles more information than they could actually carry as local hidden variables. Also, like Bell's Inequality, these inequalities must be obeyed under local realism but are violated by particles behaving according to the Copenhagen Interpretation of quantum mechanics.

With this background, let's discuss how to produce and detect entangled photons in the lab to experimentally determine who was right in the protracted debate between Einstein and Bohr on the existence of an objective reality independent of observers.

[†]This is where we have to gloss over the heavy math ...

AN ENTANGLED-PHOTON SOURCE

Before we go any further, we want to warn you that many of the experiments described in this chapter are out of the budget of many enthusiasts. This is because applications involving quantum entanglement are in the early stages of development, so the specialized crystals and detectors are not yet mass-produced. However, we feel that it is worthwhile discussing these techniques from a practical perspective, since their development is expected to lead us into a complete revolution in the way in which we compute and communicate. If the equipment is outside your budget range, you can still perform very realistic simulations using software that Mark Beck from Whitman College has kindly made available for free.[||]

At the moment, the most commonly used source of entangled particles is a crystal of β-barium borate (BBO). A single BBO crystal is able to split UV "pump" photons into so-called "signal" and "idler" photons through a process known as *parametric down-conversion*. The two down-converted photons have entangled properties, because they need to conserve the energy and momentum of their single parent photon. However, the individual properties of the photons are free to vary as long as their sum agrees with the energy and momentum of the parent photon.

There are two types of down-conversion processes that can be used in BBO crystals to produce entangled photons. As shown in Figure 139a, so-called type I down-conversion produces two down-converted photons with the same polarization, which is opposite to that of the pump photon. On the other hand, in type II down-conversion, the signal and idler photons have orthogonal polarizations, as shown in Figure 139b.

The wavelengths of the photons produced by parametric down-conversion do not necessarily have to match. In fact, the wavelengths of the down-converted photons will distribute around an average of twice the pump photon's wavelength (since on average, the down-converted photons will have one-half the energy of the pump photons). For our entangled-photon source, we will use photons that have the same wavelength $\lambda_{idler} = \lambda_{signal} = 2 \times \lambda_{pump}$.

We will use BBO crystals that support type I down-conversion, because they can be combined to produce much larger numbers of entangled photons than a type II crystal. Our entangled-photon source is based on a design by Dehlinger and Mitchell from the Physics Department of Reed College in Oregon.[45] It must be noted, however, that type II down-conversion is preferred for many applications, because it is easier to collect entangled pairs with high purity—that is, containing much lower numbers of nonentangled, stray photons.

BBO crystals cut for type I down-conversion only convert photons of a specific pump polarization. Photons of the orthogonal polarization simply pass through. For this reason, we use a design[46] by Paul Kwiat** and his colleagues at the Los Alamos National Laboratory, which calls for two identical BBO crystals, one next

---

[||]Mark Beck's Quantum Mechanics Simulations software is available for free download at: http://people. whitman.edu/~beckmk/QM/simulations/simulations.html.

**Now a professor in the Physics Department of University of Illinois.

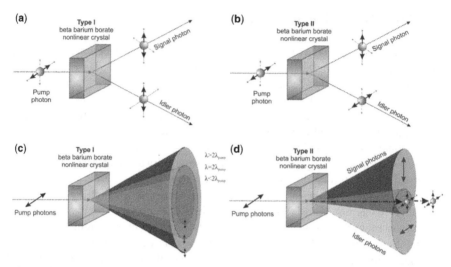

Figure 139   Nonlinear BBO crystals can be manufactured to produce entangled photons through two types of down-conversion processes. **(a)** The signal and idler photons in type I down-conversion have the same polarization, which is opposite to that of the pump photon. **(b)** Type II down-conversion produces photons with orthogonal polarization, with the idler photon having the same polarization as the pump photon. **(c)** and **(d)** The distribution of down-converted photons depends on the type of down-conversion process. Type II down-conversion produces photons with randomized polarization useful for our work only in line with the intersections between the cones.

to the other, with one of the crystals rotated by $90°$. Photop Technologies[††] can supply a mounted, ready-assembled crystal stack for around $1,400. Please note that BBO crystals are very sensitive to humidity. They must be kept in a bag with a desiccator any time they are not in use.[47]

As shown in Figure 140, a pump photon polarized at $45°$ going through the BBO crystal stack has equal probability of down-converting in either crystal. For 100-mW pump power, the rate at which polarization-entangled photons are produced can reach up to 6×10^6 entangled pairs per second. This is more than 300 times the number of entangled-photon pairs produced by type II crystals. Another advantage of this method is that the photons are entangled in energy and momentum as well. This *hyper-tangled* state is very useful when it comes to developing applications such as quantum computing, quantum encryption, and quantum communication.

High-frequency pump photons are produced by a violet $\sim$405-nm laser module. These used to cost many thousands of dollars just a few years ago, but their use in TV projectors, Blu ray disks, and laser pointers have dropped their cost to under $100. Many of these lasers produce other wavelengths besides the $\sim$405-nm line, so a band-pass filter should be placed after the laser. We use a filter made of BG3 colored glass (Thorlabs model FGB25) that allows 315- to 445-nm photons to pass through. In

---

[††]There are many companies that can supply these crystals. The current price for a stack is in the $1,000 range. A source mentioned in some of the papers is Newlight Photonics in Canada.

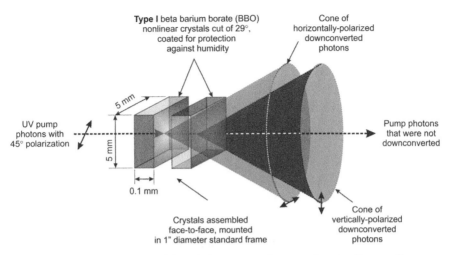

Figure 140   The nonlinear crystal in our photon entangler comprises two 5 mm × 5 mm × 0.1 mm BBO crystals mounted face-to-face at an angle of 90° to each other. Pump photons polarized at 45° produce two cones of entangled down-converted photons. The down-conversion process is very inefficient, so the vast majority of pump photons pass unimpeded by the nonlinear crystals.

addition, two washers with small holes act as baffles to keep unwanted light and reflections away. Our laser is built from a laser diode extracted from a Blu ray disk burner. We use the circuit shown in Figure 141 to drive the laser diode with 160 mA to produce around 100 mW of 405-nm polarized light. As an alternative, you may use an inexpensive (<$20) 405-nm, 5-mW laser pointer based on a Blu ray player laser diode. However, consider that the entangled-photon yield is proportional to the pump laser's power. A rough estimate for the crystals made by Photop is: number of entangled pairs per second $\approx$ pump power [mW] × 6 × 10^4.

As we discussed, type I parametric down-conversion by a BBO crystal happens only at a very specific polarization. Therefore, we need to place polarization optics between the laser's output filter and the BBO crystal. A linear glass polarizer

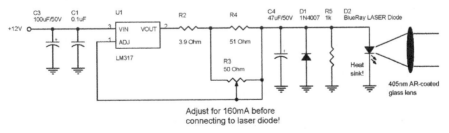

Figure 141   This 405-nm laser is based on the laser diode extracted from a Blu ray burner drive. Make sure that the laser diode is thoroughly heat-sinked, and that the current produced by the circuit (measured by replacing the laser diode with a milliammeter) is below 160 mA before connecting the laser diode.

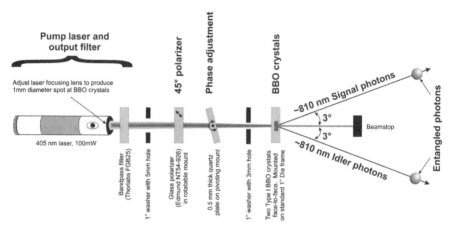

Figure 142 We use a 100-mW, 405-nm laser module to pump an assembly of two BBO crystals (Figure 140) to produce entangled IR photons at approximately twice the pump wavelength. A 45° polarizer and a thin quartz plate are used to tune the polarization and phase of the pump beam for optimal yield of entangled-photon pairs by the BBO crystals. The BBO crystals are very hygroscopic, so they must be kept in a bag with a desiccator at any time that they are not in use.

(Edmund Optics NT54-926) in a rotatable mount adjusts the polarization angle, and a small quarter-wave quartz waveplate (e.g., 0.5-mm thick x-cut polished quartz plate; Casix WPZ1310 or Edmund Optics NT83-927) is used to adjust the phase of the pump laser light to maximize entangled-photon yield. Finally, a piece of black foam is used as a beam stop for the pump laser beam that goes unimpeded through the BBO crystals. As shown in Figure 142, entangled-photon pairs of approximately the same wavelength exit the crystal at an angle of around 2.5° to 3°.

DETECTING ENTANGLED PHOTONS

Each photon in an entangled-photon pair is just a photon, and like any other photon, it can be detected using the type of methods that we used in chapter 2 (Figure 33) to detect individual photons.

However, recall that the quantum efficiency of the PMT probe that we built (Figure 30) was quite low, even at high photon energies (a maximum of 25% at 400 nm, as shown in Figure 33b). This becomes a major problem when working with entangled photons, since we want to count coincident detections when each photon is challenged in a different way in order to test their entanglement. Worse, the PMT we used is completely unresponsive to the IR photons produced by parametric down-conversion in BBO with a 405-nm pump.

Detecting individual photons at 810 nm with reasonable quantum efficiency is possible, but requires specialized detectors that are very expensive. One option is to use a low-noise PMT that is sensitive to these wavelengths—and there aren't many on the market. A possible candidate is Hamamatsu's model H7421-50 photon-counting module with a quantum efficiency of 12% at 810 nm. This module uses a metal-package PMT with a thermoelectric cooler that is necessary to reduce thermal noise generated from the photocathode. It is obviously quite a sophisticated piece of equipment, and retails at around $4,000 each (we need a minimum of two units to perform experiments with entangled-photon pairs).

A better alternative is to use a probe based on a semiconductor photodetector known as a single-photon avalanche diode (SPAD), which can reach quantum efficiencies as high as 80% at 810 nm. These detectors work somewhat like Geiger tubes (Figure 56); as you may remember from chapter 3, the gas mixture and operating voltage in a GM tube are chosen so that a single ionizing event caused by a gamma photon leads to an avalanche of ionization to produce a more easily detectable pulse. However, unlike in a GM tube, the avalanche in a SPAD (which doesn't happen in a gas, but in a solid semiconductor diode) is not extinguished (*quenched*, in the parlance of physicists) by itself, so the circuit must be designed to interrupt the current through the SPAD, readying it to detect a new photon.

As shown in Figure 143, a SPAD-based single photon–counting module (SPCM) commonly comprises a discrete SPAD packaged in a hermetic enclosure together with a thermoelectric cooler (TEC). The cooler is a module based on the Peltier Effect, which pumps heat away from the SPAD to an external heat sink when powered. The TEC is very similar to the modules used in portable beverage

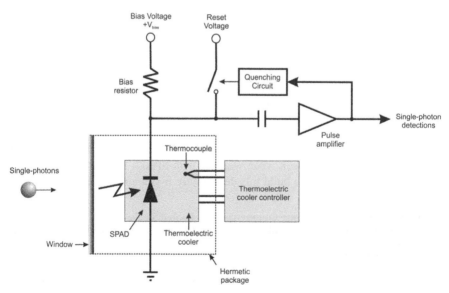

Figure 143   Modern SPCMs use SPADs as detectors. The SPAD is biased at a high voltage, which allows a single photon to cause an avalanche breakdown similar to what happens in a GM tube. However, the avalanche in the SPAD is not extinguished by itself, so a quenching circuit must be used to interrupt the current through the SPAD, readying it to detect a new photon. A thermoelectric cooler is used to keep the SPAD at a low temperature (typically 0°C to −20°C) to reduce false counts ("dark counts").

coolers. It is a simple device formed by a semiconductor sandwich that pumps heat in one direction when a DC current is applied between its terminal electrodes. A T/C sensor is commonly attached to the TEC, allowing an external circuit to control the current flowing through the TEC to maintain the SPAD's temperature at an optimal level.

The SPAD is biased through a resistor to a relatively high voltage that depends on the specific device. Typical operating voltages are in the range of 50–250 V. A photon entering the SPAD triggers an avalanche current within the SPAD that is easy to detect through an AC-coupled pulse amplifier. The avalanche current must be interrupted and the bias voltage restored before the SPAD is ready to detect a new photon. The quenching circuit to interrupt the avalanche current may be implemented through simple passive biasing, using correctly matched resistors, or through an active mechanism that forcibly turns off and then restores the bias to the diode when a photon is detected. In general, active circuits are much more complex, but allow the SPAD to detect consecutive photons at rates that passive quenching would miss. Passive quenching is now more commonly being applied in newer designs, and some of the newer devices even integrate a quenching mechanism embedded within the SPAD, providing effective feedback for stabilization and quenching of the avalanche process without any external support circuitry.

There are very few SPAD manufacturers. The most notable are PerkinElmer (as part of their Excelitas Technologies division), and Laser Components. However, applications for SPADs are expanding, particularly in biomedical applications such as DNA sequencing, so their availability and pricing may eventually bring them within range of a larger number of physics experimenters. Table 9 shows some of the currently available commercial SPADs that can be used to detect entangled photons at wavelengths close to 810 nm.

Figure 144 shows the schematic diagram for a passively quenched SPCM based on a Perkin-Elmer C30902S-DTC SPAD. Besides our design, you may also want to take a look at the high-performance, relatively low-cost design for an actively quenched SPCM based on this same SPAD that was undertaken as a senior design project at the University of Illinois. Their report, including complete schematics, is available online.[65] Please note that gaining sufficient experience to build a high-performance SPCM is difficult when working with SPADs that cost around $1,000 apiece, so you should try to build this device only if you are very familiar with the

TABLE 9   Characteristics of Some Commercial TEC-Cooled Single-Photon Avalanche Photodiodes Suitable for Experiments with Entangled Photons at an Approximate Wavelength of 810 nm

Manufacturer	Model	Quantum efficiency at 810 nm	Dark count rate [per second]	Approximate unit cost ($)
Perkin-Elmer	C30902SH-TC	77%	1,100	700
Perkin-Elmer	C30902SH-DTC	77%	250	1,050
Laser Components	SAP500T6	60%	5,000	650
Laser Components	SAP500T8	60%	2,000	1,000

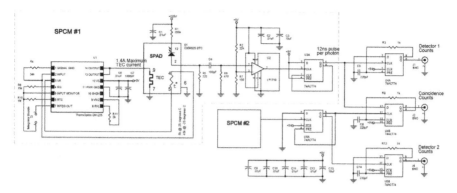

Figure 144   Two TEC-cooled SPADs are used to build two SPCMs. The SPCM outputs
are processed by two 74ACT74 dual D-type positive edge-triggered flip-flops that
perform pulse-stretching and coincidence-counting functions for Bell's Inequality tests.

construction of low-noise electronic systems that incorporate high-speed logic.
Acquiring ready-made SPCMs may sometimes be a better alternative (e.g., Perkin-
Elmer SPCM-AQRH at around $2,600 each for the 250 dark counts/s version, or
the 1,100 dark counts/s version for $2,250).

In the circuit of Figure 144, the SPAD is reverse-biased through a 200-kΩ resistor. This value is sufficiently large that an avalanche in the SPAD will be quenched by
itself within less than a nanosecond. After quenching, the voltage across the SPAD
recovers toward its original working bias voltage following an exponential increase
with a time constant of around 400 ns. The SPAD does work while the bias is recov-
ered, but at much lower photon-detection efficiency than at nominal bias voltage. The
pulses produced by the SPAD are AC-coupled through C4 to a fast, constant-level discriminator built from a LT1719 fast comparator. The threshold level for the comparator is fixed at 22.5 mV. The LT1719 has an output that is compatible with TTL logic
circuits. One of the dual D-type positive edge-triggered flip-flops inside a 74ACT74
IC shapes the comparator's output into a 12-ns pulse every time a photon is detected.

The SPAD is cooled by a TEC module that is built into the SPAD's package. The
TEC controller is a dedicated circuit by ThermOptics that implements a feedback loop.
The DN1225 operates from a single 5-V power source and is capable of supplying 2 A
of current to a 2-Ω load with better than 90% efficiency. The device interfaces directly
to the negative temperature coefficient resistor within the C30902S-DTC that senses
the SPAD's temperature. The temperature setpoint is selected via resistor R4 to cool
the SPAD to $-20°C$. R8 sets the proportional gain to 500, and R10 sets the integrator
time constant to 1 s. Finally, R11 sets the maximum voltage supplied to the TEC to 1 V.
The DN1225's proportional-integral control loop is capable of achieving a temperature
stability of 0.01°C. The DN1225 is not always available, but ThermOptics' DN1221
can also be used if the circuit is adapted to this module's pinout.

The second SPCM is identical to the circuit we just described. The SPCM outputs are processed by a coincidence-counting circuit designed by Dehlinger and
Mitchell from Reed College.[45] Output pulses are 250-ns long, and should be fed to
a three-channel counter. One possibility is using a PC-based digital timing/counting

(a)

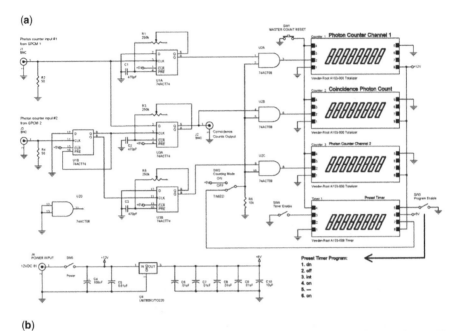

(b)

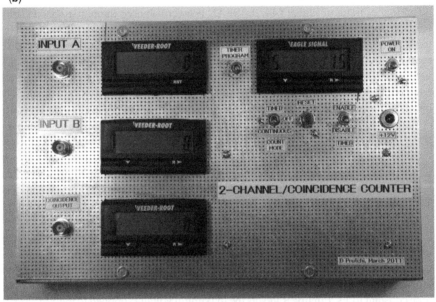

Figure 145   We use this two-channel photon and coincidence counter to monitor and tune our setup. **(a)** The SPCM outputs are processed by two 74ACT74 dual D-type positive edge-triggered flip-flops that perform pulse stretching and coincidence counting. The pulse width into each Veeder-Root A103-000 counter should be adjusted to 15 μs. The A103-008 allows counting to happen for a preset amount of time. **(b)** The circuit fits neatly within an 11-in. × 7-in. × 2-in. aluminum chassis. We used a sheet of breadboard material for the panel, because it makes it easy to cut rectangular holes for the Veeder-Root modules.

board (e.g., a PCI-6601 by National Instruments). Alternatively, you could feed the SPCMs' outputs directly to the stand-alone coincidence counter shown in Figure 145 to test the CHSH version of Bell's Inequality.

HIGH-PURITY SINGLE-PHOTON SOURCE

Our single-photon experiments (Figure 33, Figure 90, Figure 121, Figure 125, and Figure 132) have all used a highly attenuated laser beam to produce single photons. However, a precise analysis of this method shows that we could be assured that single photons fly through the apparatus only if the source photons are completely independent of each other. But photons sometimes bunch together, allowing pairs and triplets to fly together through the system, resulting in data that does not pass the rigorous scrutiny required for cutting-edge research.

An entangled-photon setup, such as the one shown in Figure 146, is now commonly used in high-level research involving single photons, because the down-conversion process does not produce bunched photons. In addition, one of the two entangled photons is used to "gate" the detection system. Single photons detected

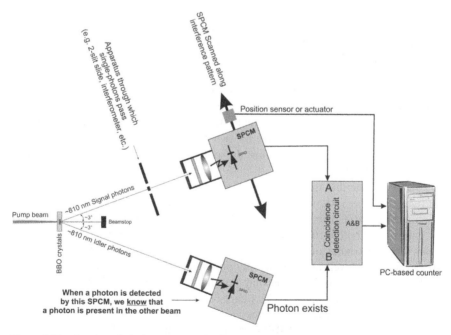

Figure 146 An entangled-photon source is often used in advanced single-photon experiments, because the down-conversion process intrinsically produces two beams of individual photons that verify each other's existence. A photon reaching the experiment's detector (channel A) is counted only if it coincides with the detection of its partner photon (channel B). This eliminates the possibility of "photon bunching," which may affect single-photon sources based on strongly attenuated laser beams.

after they fly through the experimental apparatus (e.g., double slit, Mach–Zehnder interferometer, etc.) are counted only if their arrival is expected as announced by the gating photon. This technique considerably reduces background counts due to detector noise or stray photons.

Other single-photon sources are being investigated because of their importance for quantum technology (e.g., in quantum cryptography), as we will see a bit later. Most of these rely on tightly focusing a laser beam onto a sample area containing a very low concentration of quantum dots, such that only one quantum dot becomes excited and emits a photon at any given time. These sources are inherently anti-bunched,[68] and can produce high-intensity beams of single photons at visible wavelengths.

TESTING BELL'S INEQUALITY

In 1982, French physicist Alain Aspect was able to run the first experimental test of Bell's Inequality.[50] Aspect used a very complex entangled source in which calcium atoms (in an atomic jet) are pumped by a two separate lasers.[51] Using the CHSH form of Bell's Inequality, Aspect was looking for the value of a certain coincidence statistical parameter (S) to be equal or lower than two if local hidden variables are present. On the other hand, quantum mechanics predicted $S = 2.7$. Aspect's experimental result was $S = 2.697 \pm 0.015$, providing overwhelming support to the thesis that Bell's Inequality is violated in its CHSH version, thus confirming quantum indeterminacy.

Figure 147 and Figure 148 show how you can use the entangled-photon source that uses two BBO crystals (Figure 142) and SPAD-based SPCMs (Figure 144) to conduct a modern replication of Aspect's experiment. The general idea is to count photons detected by two SPCMs, as well as their coincidence counts[‡‡] for the set of angles for the polarizers that precede the SPCMs, as shown in Table 10.

Please note that the detectors are very sensitive to photons at the UV pump wavelength. The 780-nm long-pass filters drastically reduce the number of UV photons that make it through, but the UV photons would far exceed the number of entangled photons if shading were not used. We constructed the combined beam stop/collimator of Figure 149 to absorb pump photons that go through the BBO crystals, as well as to allow only photons in the $\pm 3°$ beams to pass through to the detecors. If required, you may install an IR long-pass filter (e.g., Thorlabs FGL15S or a photographic R-72 filter) behind the collimation holes to further reduce UV photon leakage. Another trick that works well is to place the laser relatively far away from the BBO crystals (you may bend the path around with

---

[‡‡]Example LabView programs written by Mark Beck of Whitman College for this specific purpose can be downloaded from http://people.whitman.edu/~beckmk/QM/labview/labview.html. These programs are designed to control motorized polarizer mounts and other actuators, and they will not run without the specific hardware configuration used at Whitman College. However, they make a great guide to writing programs to suit your own setup.

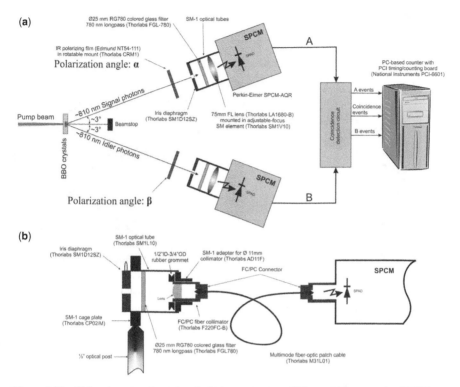

Figure 147   This setup uses the entangled-photon source of Figure 142 to test the CHSH version of Bell's Inequality. The coincidence detection circuit is shown in Figure 144. **(a)** In its simplest form, photons are filtered and focused directly onto the SPADs. **(b)** We use Perkin-Elmer SPCMs with fiber-optic connector inputs, which allow us to route photons from small, light collection heads to the SPCMs placed outside the optical breadboard.

UV-enhanced aluminum mirrors if necessary) such that the laser's spot at the beam stop is small, and does not overlap with the $\pm 3°$ entangled-photon beams.

Before taking measurements to test Bell's Inequality, your setup will need to be aligned[§§] to yield the highest count of entangled photons on the coincidence channel when the polarizers in front of the SPCMs are set to the same polarization angle ($\alpha = \beta$). This requires an enormous amount of patience. It is best accomplished by first aligning the major mechanical components (without the IR filters) using a visible laser projected "backward"—that is, starting from the position that the SPCMs will occupy, and then toward the plane where the BBO crystals will lie (we substitute the BBO crystals by a 1-in.-OD $\times \frac{1}{4}$-in.-ID steel washer). We use a

---

[§§]Enrique Galvez from Colgate University wrote a detailed guide to performing experiments using entangled photons with a substantially similar setup. It is titled "Correlated-Photon Experiments for Undergraduate Labs," and is available for free download at http://departments.colgate.edu/physics/research/Photon/Photon%20research/Quantumlan07/Lab%20Manual2010.pdf.

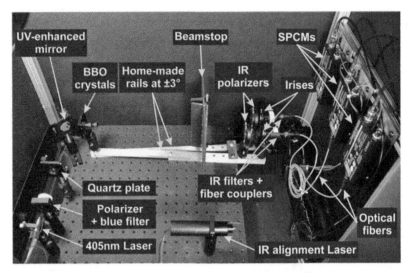

Figure 148 Our entangled-photon setup is enclosed within a light-tight box made of black foam board. The laser, BBO crystal, and aluminum arms holding the fiber-optic collectors fit on an 18-in. × 18-in. bench plate. Fiber-optic jumpers convey photons to three surplus Perkin-Elmer SPCMs that we purchased on eBay.

TABLE 10 Photon Counts and Coincidence Counts you should Acquire for these Polarization Settings for the Polarizers Preceding the SPCMs of Figure 147

Case	Polarization angle α for polarizer preceding SPCM "A"	Polarization angle β for polarizer preceding SPCM "B"
1	−45°	−22.5°
2	−45°	22.5°
3	−45°	67.5°
4	−45°	112.5°
5	0°	−22.5°
6	0°	22.5°
7	0°	67.5°
8	0°	112.5°
9	45°	−22.5°
10	45°	22.5°
11	45°	67.5°
12	45°	112.5°
13	90°	−22.5°
14	90°	22.5°
15	90°	67.5°
16	90°	112.5°

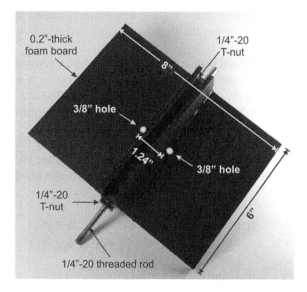

0.2"-thick foam board

1/4"-20 T-nut

8"

3/8" hole

1.24"

3/8" hole

1/4"-20 T-nut

6"

1/4"-20 threaded rod

Figure 149 Many UV photons can go through the 780-nm long-pass filters preceding the detectors, so we constructed this beam stop to shade the detectors from photons that don't belong to the entangled beams at $\pm 3°$. The foam board is simply hot-glued to two $\frac{1}{4}$-in. T-nuts positioned on a $\frac{1}{4}$-in. 20-threaded rod. This beam stop is screwed on the breadboard 12 in. away from the BBO crystals.

red laser pointer that we focus onto a Thorlabs F220FC-B fiber coupler as a laser source for this step. We can therefore project a laser from the position of the photon-collection heads (Figure 147b) by injecting the laser into the fiber-optic patch cords. Next, we place a mirror within the BBO crystal holder and further refine the alignment by checking that the reflection of the laser projected from one of the photon-collection assemblies reaches the other photon collector. At this point, we place the BBO crystal and IR filters in place, substitute the red laser with an IR unit (30 mW, 980 nm; Information Unlimited model LM980-30), and tweak everything again before connecting the fiber-optic patch cords to the SPCMs.

Next, we align the 405-nm pump laser. This laser should be rotated to maximize the intensity of the beam going through the 45° glass polarizer placed before the BBO crystals. The next step is to find the best angle for the SPCMs. Turn off all lights and turn on the SPCMs. Turn on the pump laser and move one of the SPCMs around $3° \pm 1°$ to find the angle that produces the most counts. Set the second SPCM to the same angle. Next, ensuring that $\alpha = \beta = 0°$, move the SPCMs very lightly to maximize the number of coincidence counts. Depending on your laser's power, the number of coincidence counts should be in the range of 300–1,000 counts per second when the irises on the SPCMs are fully opened. Next, adjust the quartz $\frac{1}{4}$-wave waveplate to maximize the number of individual and coincidence counts. Then, change the polarizing angles $\alpha = \beta = 45°$, and tweak the pump laser's polarization so the coincidence counts obtained at $\alpha = \beta = 45°$ are as close as possible to those at $\alpha = \beta = 0°$. Lastly, substitute a 1-in.-OD $\times \frac{1}{4}$-in.-ID steel washer for the BBO crystal assembly temporarily, turn on the pump laser, and obtain channel and coincidence-count background for your system at all angle settings (all angles shown in Table 10). You will need to subtract these background counts from your raw data to estimate the number of events caused by entangled IR photons. We must stress again that the alignment process is very tedious and demands great patience!

We are finally ready to collect actual data at the polarization angles given in Table 10. Run the acquisition until you get at least 1,000 coincidence counts (after subtracting background counts) for the ($\alpha = -45°$, $\beta = -22.5°$) setting. The exact amount of time is not important, but the same acquisition time should be used for all settings.

Testing the CHSH version of Bell's Inequality involves calculating the *correlations* between coincidence counts at the various angle settings. Please take a look at Table 11, where the correlations are calculated for the 16 cases using our setup:

$$E(\alpha, \beta) = \frac{N(\alpha, \beta) + N(\alpha + 90°, \beta + 90°) - N(\alpha + 90°, \beta) - N(\alpha, \beta + 90°)}{N(\alpha, \beta) + N(\alpha + 90°, \beta + 90°) + N(\alpha + 90°, \beta) + N(\alpha, \beta + 90°)}$$

For example, E for case 1 is calculated:

$$E_1 = E(-45°, -22.5°)$$

$$= \frac{N(-45°, -22.5°) + N(45°, 67.5°) - N(45°, -22.5°) - N(-45°, 67.5°)}{N(-45°, -22.5°) + N(45°, 67.5°) + N(45°, -22.5°) + N(-45°, 67.5°)}$$

$$= \frac{1,079 + 1,043 - 322 - 410}{1,079 + 1,043 + 322 + 410} = \frac{1,390}{2,854} = 0.487$$

As you can see from Table 11, there are really only four unique correlations for this set of 16 angle combinations. We have bolded them on the table. The first is the one that we just calculated. The other three are:

$$E_2 = E(-45°, 22.5°)$$
$$E_3 = E(0°, -22.5°)$$
$$E_4 = E(0°, 22.5°)$$

Let's skip over the derivation[52] of the CHSH form[¶¶] of the Bell Inequality for our setup, and jump straight to the inequality we need to test. The quantity to calculate is:

$$S = E_1 - E_2 + E_3 + E_4$$

[¶¶]However, this is exactly how the CHSH inequality is applied in experiments. The CHSH inequality is $-2 \leq S \leq 2$, where $S = E(a, b) - E(a, b') + E(a', b) + E(a', b')$. a and a' are detector settings on side A, and b and b' are detector settings on side B, which are the four combinations being tested in separate sub-experiments. The terms $E(a, b)$, etc. are the quantum correlations of the particle pairs.

TABLE 11   Coincidence Counts and Calculation of Correlations to Test the CHSH Version of Bell's Inequality[a]

Case	Polarization angle α for polarizer preceding SPCM "A"	Polarization angle β for polarizer preceding SPCM "B"	Coincidence counts (N)	Correlation $E(\alpha, \beta)$
1	$-45°$	$-22.5°$	1,079	$E_1 = 0.487$
2	$-45°$	$22.5°$	268	$E_2 = -0.591$
3	$-45°$	$67.5°$	410	-0.487
4	$-45°$	$112.5°$	1,074	0.591
5	$0°$	$-22.5°$	1,170	$E_3 = 0.701$
6	$0°$	$22.5°$	1,110	$E_4 = 0.527$
7	$0°$	$67.5°$	212	-0.701
8	$0°$	$112.5°$	358	-0.527
9	$45°$	$-22.5°$	322	-0.487
10	$45°$	$22.5°$	1,074	0.591
11	$45°$	$67.5°$	1,043	0.487
12	$45°$	$112.5°$	285	-0.591
13	$90°$	$-22.5°$	216	-0.701
14	$90°$	$22.5°$	330	-0.527
15	$90°$	$67.5°$	1,261	0.701
16	$90°$	$112.5°$	1,108	0.527

[a]Data collected using the setup of Figure 147. The four unique correlations in bold are used to calculate the quantity S that determines whether Bell's Inequality is respected or violated.

where $|S| \leq 2$ if local hidden variables exist, and $S > 2$ if the Copenhagen Interpretation is correct.[‖‖] For the experimental data in Table 11, $S = 2.305$, which is a clear violation of the Bell Inequality!

CLOSING THE LOOPHOLES

There are a number of experimental problems or "loopholes" that may affect the validity of results in Bell test experiments conducted with the type of apparatus we just described. The main challenge relates to the low detection efficiency of optical systems, and the way in which it affects the "fair sampling" of coincidences that could bias the results in favor of quantum entanglement rather than local hidden variables.

We must recognize that the detection efficiency of our optical system is much lower than the $\sim 70\%$ quantum efficiency of the SPADs. A huge number of photons are lost along the way, especially within the polarizers, lenses, and filters, as well as when photons cross the interface between these optical components and air. Our detection efficiency with the setup of Figure 147 is probably around 5%. The best optical experiments conducted with high-quality components and using fiber-optics may approach a detection efficiency of up to 35%.

‖‖Or, if you prefer, at least that the photons are truly entangled, as predicted by quantum mechanics, without the need for local hidden variables.

Much theoretical work has been done to determine the effect of low-efficiency detection, and the consensus is that a sufficiently large sample of detected pairs is representative of the pairs emitted, irrespective of the efficiency. Unfortunately, there is no way to experimentally test whether a given experiment does fair sampling, and that is why we recommend extending the sampling time to collect at least 1,000 coincident counts for case 1.

There are some Bell test experiments that have been performed using two ions rather than photons, which allowed the detection efficiency to be 100%, thus closing the fair sampling loophole. For example, in 2001, physicists of the U. S. National Institute of Standards and Technology and the University of Michigan[53] conducted a Bell test experiment using entangled beryllium ions ($^9Be^+$). The test resulted in $S = 2.25 \pm 0.03$, which clearly violates the $S \leq 2$ limit for local hidden variables.

Another issue with the experiment we conducted is that it does not exclude the possibility that some unknown communication mechanism would be available to the photons that would allow the correlations to be higher than the classical limit of $S \leq 2$. John Bell proposed that a way of closing this loophole would be to choose the settings of the polarizers only after the photons had left the source. Alain Aspect thus performed an additional experiment[54] in which the polarization was changed very rapidly within the experiment, ensuring that detected photons had already left the source before the polarization state was set. Since then, ever more sophisticated experiments—including experiments by Nicolas Gisin's group at the University of Geneva[55] using detectors 18 km apart—have confirmed that there is indeed a strange "quantum connectedness" between the two particles in an EPR experiment. Somehow each "knows" what is happening with the other.

One last important thing to mention is that, although quantum physics shows that entangled particles in different places remain "connected" regardless of how much time or space separates them, this "spooky action at a distance" does not violate the Theory of Relativity, because it is impossible to use the effect to send information.

This requires a bit of explanation: the stream of counts coming out of only one of our detectors in Figure 147 will be just a random sequence, regardless of the orientations of the polarizers. That is, we cannot change around the polarizer on the other detector to send a message via entanglement. Any information encoded in the entanglement is only extractable when you look at *correlations* between measurements on both the entangled photons. However, to access that correlation information, you need conventional (i.e., *not* faster than light) communication between the detectors and the coincidence counter.

THE AGE OF QUANTUM INFORMATION

So far, we have explored the fundamentals of quantum physics, perhaps paying a bit too much attention to its mind-boggling philosophical implications. In the real world of academic and industrial physics however, quantum mechanics is applied very successfully to the solution of physics problems and in the development of electronic devices without the need for a philosophical interpretation.

When pushed for a philosophical stance, professional physicists often fall back on the expression "shut up and calculate," which is often attributed to Nobel Prize recipient Richard Feynman. This is because, despite all the business about Heisenberg's Uncertainty, wave–particle duality, quantum entanglement, and the counterintuitive nature of quantum behavior, quantum physics nevertheless makes exact predictions that are useful for the development of real-world applications. Semiconductors, superconductors, lasers, magnetic resonance imaging machines, and other many of our modern everyday devices involve an understanding and application of the principles of quantum physics. Each of these has matured into its own field.

In this final section, we will distance ourselves from philosophy, and instead take a brief look at the way in which physicists are beginning to harness quantum weirdness to develop new technologies that will take us from the information age into the *quantum information* age.

Currently, the basic unit of digital information is the bit, which can have a value of either zero or one. In a computer, these digital values may be represented by two different voltages. For example, in a TTL circuit, "0" = 0 V and "1" = 5 V. When communicating through a fiber-optic cable, the bit's value could be encoded through two different light polarizations. For example, "0" = vertical polarization and "1" = horizontal polarization.

These orthogonal polarizations could be easily analyzed through a special type of beam splitter that separates a beam of light depending on its polarization. As shown in Figure 150, a *two-channel beam splitter* or *polarizing beam splitter* (usually called a *PBS* for short) transmits all photons with vertical polarization, but reflects all those

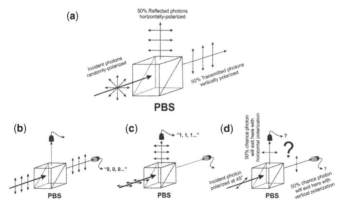

Figure 150 A PBS transmits all photons with vertical polarization, but reflects all those with horizontal polarization. (**a**) Randomly polarized light is divided into two equal beams, one vertically polarized and one horizontally polarized. Digital information can be encoded on a light beam by changing its polarization, such that a PBS can decode vertical polarization as a digital "0" (**b**), and horizontal polarization as a "1" (**c**). (**d**) However, a photon polarized at an intermediate angle will be in quantum superposition. Each such photon may be detected exiting one or the other port, with probability dependent on the polarization angle. Each photon in quantum superposition can then be thought of as a quantum bit or *qubit*.

with horizontal polarization. However, a photon polarized at any intermediate angle between horizontal and vertical will be in quantum superposition. Each such photon may be detected exiting one or the other port, with probability dependent on the polarization angle.

So long as it is not detected, a photon in a superposition of states could thus be considered to be encoding a bit in both the vertical and horizontal polarizations at the same time. This is the basic unit of data in the quantum world, and is known as a *qubit* (for *quantum bit*).

A QUANTUM RANDOM-NUMBER GENERATOR

It isn't easy at all to understand how a bit that is both a digital one and a digital zero at the same time could be used to convey or process data. In fact, the perfectly random outcome of measuring such a bit is now used to generate random streams of numbers.

The random-number function in your computer produces only a random-looking sequence of numbers through a specific algorithm. If you know the algorithm and the position in the sequence, you can accurately predict what will be the next number in the sequence. While this may be good enough for home computer games, a wide variety of high-security applications—from online casinos to jamming enemy communications—require truly random sequences of numbers that cannot be guessed through any means.

Traditionally, true random-number generators have relied on measuring events produced by random processes, such as thermal noise in electronic components or the timing between radioactive decays of an isotope. However, the measurement equipment itself somewhat biases the random-number sequence, which allows a level of prediction by very sophisticated users, opening up vulnerabilities in very sensitive applications.

Quantum random-number generators overcome these biases and are now commercially available for use in critical applications. One popular scheme is shown in Figure 151. Here, a photon polarized at a 45° angle is in quantum superposition, and will be detected in one or the other port of the PBS with 50%/50% probability. The 45° photons are produced by an entangled-photon source, but the entanglement in itself is not used. Instead, the idler photon, which is generated at exactly the same time as the signal photon, is used to prompt the system that an event detected by the SPCMs looking at the outputs of the PBS is the result of an actual 45° signal photon, and not the result of a stray photon or dark count.*** In 1999, Thomas Jennewein from Anton Zeilinger's group in Austria demonstrated that no prior method has produced a higher-quality random-number sequence than the one produced in this manner.[56]

***Commercial quantum random number generators also correct for coincidental counts that may happen because two detectors produce a dark count at the same time. In addition, supplementary optics are used to ensure as perfect a 50%/50% splitting by the PBS as possible.

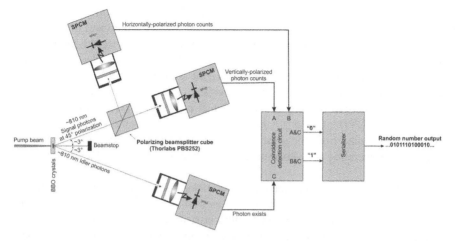

Figure 151   A popular quantum random-number generator produces a true random sequence of digital ones and zeros by detecting photons polarized at 45° through a PBS. The photons are produced by an entangled-photon source, although the entanglement in itself is not used. Instead, the idler photon is used to prompt the system that an event detected by the SPCMs looking at the outputs of the PBS is the result of an actual 45° signal photon, and not the result of a stray photon or dark count.

QUANTUM INFORMATION

Let's try to shed some light onto the matter of quantum information by defining "information" as something that is encoded in the state of a physical system. As we have discussed, information may be encoded in the voltage on a line, a specific polarization of light, the side of a coin facing up, or vibrations in the air that we decode as a spoken word. In the case of the quantum world, John Bell showed that quantum information is encoded in *nonlocal correlations* between the different parts of a physical system. As we know by now—and probably the reason why this is all so confusing—is that these correlations have no classical counterpart. We can thus expect that quantum information will demonstrate many unusual properties.

The first difference is that while classical information is independent of its physical representation, quantum information cannot be read without disturbing the physical process that encodes it. For example, the number "9" in binary is "1001" and could be transmitted through a wire encoded as the series of voltages "5 V, 0 V, 0 V, 5 V." It could be measured with an oscilloscope, and then written on the back of a napkin as "9", "nine", "nueve", "1001", and so on without changing the actual information that is encoded. These are only representations of the number "9," and measuring it with an oscilloscope or writing it down on a piece of paper does not prevent a computer down the line from correctly decoding the number "9." In contrast, it is impossible to write down on paper the previously unknown information contained in the polarization of an entangled-photon pair.

The second difference stems from Heisenberg's Uncertainty Principle. We know that measuring any property of a quantum system introduces a disturbance, making it impossible to copy quantum information with perfect fidelity (this is known as the *quantum no-cloning principle*). If we could make a perfect copy of a quantum state, we could measure a property of the copy without disturbing the original and we could thus defeat Heisenberg's Uncertainty Principle. On the other hand, nothing prevents us from making as many perfect copies of classical information. For example, our DVD of *Back to the Future* contains the same exact bits as every other DVD printed from the same master. In addition, the witty plot and outcome don't vary depending on who watches the movie, or whether it has been watched before.[†††]

These properties make quantum information sound like an unnecessarily difficult way of encoding information that we, as well as our computers and telecommunication networks, handle very well with classical information. However, the very peculiar nature of quantum physics enables many tricks that are simply impossible with classical communications and computing. For example, the fact that measuring a quantum system causes it to collapse into a certain state has been used to build secure communication channels that show with certainty if someone has eavesdropped.

QUANTUM TELEPORTATION

It would seem that it is impossible to communicate or manipulate an unknown quantum state without destroying its fragile nature and causing its collapse. Furthermore, we have already seen that any measurement, however mild, causes Heisenberg's Uncertainty Principle to kick in and cause a random change in the system being measured.

In 1993, scientists Charles H. Bennett of IBM; Gilles Brassard, Claude Crépeau, and Richard Jozsa from the University of Montreal; Asher Peres from the Technion in Israel; and William K. Wootters of Williams College proposed that entanglement could be used to get around the limitations imposed by Heisenberg's Uncertainty Principle in order to communicate an unknown quantum state.[57] Their proposal was to *teleport*[‡‡‡] the quantum state of a particle from one place to another through the use of an auxiliary pair of entangled particles and a classical communications channel.

The idea, shown in the diagram of Figure 152, is as follows: the transmitter—commonly referred to as *Alice*—is given a photon with unknown state *x* to be teleported to a receiver—*Bob*—without either of them learning anything about *x*'s state. Expecting this task, Alice and Bob previously shared an entangled pair of photons, which they have kept undisturbed. Alice performs a joint measurement of her entangled photon A and the photon to be teleported (*x*). However, this joint

---

[†††]Something that we can say for certain, after conducting this experiment at least 32 times.

[‡‡‡]A name inspired by the *Star Trek* method of transporting crew members to a planet's surface. Funnily enough, *Star Trek* creator Gene Roddenberry conceived of teleportation as a way of cost-cutting because simulating landings and takeoffs of shuttle craft was very expensive.

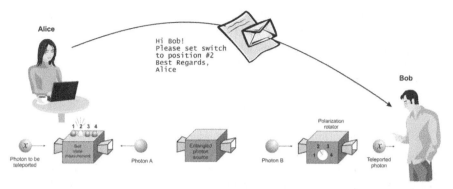

Figure 152   The unknown state of photon x can be teleported by Alice to Bob if they share an auxiliary entangled-photon pair. Alice first performs a Bell-state measurement between her entangled photon (A) and the photon to be teleported (x). Bob's photon instantaneously acquires a state that incorporates the unknown state of x. Alice uses a classical communication channel to tell Bob how to rotate the polarization of his photon based on the results of the Bell-state measurement to produce a perfect replica of photon x. However, the state of x remains unknown, thus achieving teleportation.

measurement does not allow Alice to determine the individual polarizations of either A or x, so it is not bound by Heisenberg's Uncertainty Principle. Instead, it tells Alice only what the relative polarization between A and x is. For example, Alice may find that A and x have the same polarization, or that their polarizations are perpendicular to each other, or that they are at an intermediate angle. This measurement, known as a *Bell-state measurement*, destroys x, but not the quantum connection between A and B. In fact, Bob's photon (B) instantaneously acquires a state that incorporates the unknown state of x. The teleportation is completed when Alice tells Bob how to rotate the polarization of his photon to remove the original state of the entangled pair, leaving Bob with an exact replica of the original photon x. Again, neither Alice nor Bob know anything about x's state, allowing it to remain in quantum superposition.

The last procedure in the teleportation sequence requires Alice to tell Bob the results of her Bell-state measurement via a classical channel (e.g., a letter, e-mail message, or phone call). Only after receiving this information via a normal communications channel (which does not exceed the speed of light), can Bob potentially read and use the polarization state that was conveyed instantaneously via "spooky action at a distance" between the entangled pair of photons, thus preserving Einstein's limit on the speed at which information can be passed between two observers.

Let's take a look at the teleportation protocol in more detail. Suppose that photon x is a qubit in superposition of states:

$$|\Psi\rangle_x = a|\text{vertical}\rangle_x + b|\text{horizontal}\rangle_x = a|0\rangle_x + b|1\rangle_x$$

where the bra-ket notation $|\psi\rangle_n$ simply denotes that the system or particle n has the state ψ.

At the same time, Alice and Bob have a pair of entangled photons A and B. Assuming that we use a type II BBO crystal to generate the entangled pair, the entanglement places both photons into a single quantum system AB, where both photons have unknown polarization, but their individual polarization is perfectly perpendicular to one another:

$$|\Psi\rangle_A = \text{perpendicular to } |\Psi\rangle_B$$

When Alice performs the Bell-state measurement between photons x and A, the entanglement originally shared between Alice's and Bob's photons is broken. Instead, Bob's photon B now takes on a state $|\Psi\rangle_B$ that is a superposition of four possible states that combine the original state of x with the original state of A. One of these four random states is already the original state of x, the other three are states in which the polarization of x has been rotated by a certain amount:

1. $|\Psi\rangle_x = |\Psi\rangle_B$, the state of Bob's photon is already identical to the original state of x.

2. $|\Psi\rangle_x = |\Psi\rangle_B + 90°$, the state of Bob's photon is perpendicular to the original state of x.

3. $|\Psi\rangle_x = |\Psi\rangle_B + \alpha$, the state of Bob's photon is the original state of x rotated by some angle α.

4. $|\Psi\rangle_x = |\Psi\rangle_B + \beta$, the state of Bob's photon is the original state of x rotated by some angle β.

The last two states involve a complex rotation, so let's not worry about them. They won't prevent us from understanding the process.

Alice knows which one of the polarization rotations has to be applied by Bob to recreate the original state of x, but Bob doesn't know it until he receives a conventional message from Alice. Once Bob knows the results of Alice's Bell-state measurement, he simply uses a polarization rotator (which doesn't tell him the actual polarization of the photon, it simply adds a certain polarization angle to whatever polarization the photon already had) to perfectly reproduce the state of x.

So, if Alice's Bell-state measurement in Figure 152 results in case 1, she tells Bob that his photon is already identical to the original state of x. However, if Alice measures case 2, she needs to tell Bob to rotate his photon by 90° to recreate the original state of x. At the end of the procedure, the original photon x at Alice's end has been destroyed, but its state has been teleported with complete precision to a photon at Bob's end. Since the photon is completely characterized by its quantum state, teleporting the state is identical to actually teleporting the photon itself.

Bennett et al.'s 1993 teleportation protocol looks simple,[57] but it took the brilliance of Anton Zeilinger—arguably today's most prominent experimental quantum physicist—to figure out how to implement it in the real world. It was only in 1997 that Anton Zeilinger's team, then at the University of Innsbruck in Austria, was able to perform an actual teleportation experiment with the setup shown in Figure 153. This diagram is rather busy, so let's take our time looking at each part of the system.

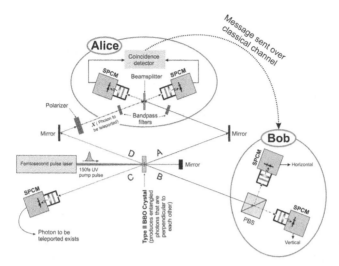

Figure 153   In the system developed to perform the first successful teleportation, an entangled pair of photons A and B is produced by a very brief (150×10^{-15} s) laser pulse. One photon from this pair goes to Alice and the second to Bob. The pump pulse is then reflected back into the BBO crystal to produce a second set of entangled photons C and D, but one of them (D) is given a known polarization to create the photon x to be teleported. The second photon (C) is used to indicate the existence of x. Alice conducts a Bell-state measurement between her entangled photon (A) and x (for which she doesn't know the polarization). Alice tells Bob when both of her detectors fire simultaneously. When this happens, Bob verifies that his entangled photon (B) has the same polarization as x, thus demonstrating successful teleportation.

The process begins with a pulse of UV light produced by a very specialized (and currently very expensive) type of laser.[§§§] The laser's pulse lasts only 150 fs (150×10^{-15} s), which barely suffices to produce one entangled-photon pair (A and B) when sent through a BBO crystal. Instead of dumping the remaining UV light into a beam stop, as we did before, the pulse is reflected back by a mirror into the BBO crystal to produce a second set of entangled photons C and D very shortly after A and B are formed.

A and B are shared between Alice and Bob as the auxiliary entangled photons required for teleportation. Photon D is passed through a polarizer to produce a photon to be teleported (x) for which we—but importantly neither Alice nor Bob—know the polarization. Passing D through a polarizer destroys its entanglement with C, but this is not important, because C is used only to tell the system that x is available for teleportation.

So, passing the brief laser pulse through the nonlinear crystal has given us four almost simultaneous photons: the pair that remains entangled (A and B) are sent to

---

[§§§]The pump laser that is commonly used is based on a mode-locked Ti:sapphire laser (e.g., Coherent Mira 900-F) pumped by an argon laser (e.g., Coherent INNOVA Sabre). The output is a train of infrared (790 nm) 120-fs pulses at a rate of 76 MHz. The infrared pulses are converted to ultraviolet using a lithium triborate (LBO) nonlinear crystal cut for second harmonic generation. A typical conversion efficiency from IR to UV is about 40%, which yields a 300-mW beam of UV pulses shorter than 200 fs.

Alice and Bob. The other pair is used to create x and a trigger signal for the teleportation system.

The Bell-state measurement proved to be the most difficult challenge for Zeilinger's team. The idea is that the two photons to be measured jointly must arrive at exactly the same time to a beam splitter. The photons have three ways of exiting the beam splitter: they can both exit through one port, they can both exit through the other port, or they can each exit through separate ports.

We won't go into the mathematics of the Bell-state measurement, but suffice it to say that the photons exit through different beam-splitter ports only 25% of the time, and when this happens the Bell state measurement indicates that $|\Psi\rangle_x = |\Psi\rangle_B$. That is, the state of Bob's photon has been converted to be identical to the original state of x. The other 75% of the time, photons exit together through one or the other port of the beam splitter, indicating that some rotation has to be applied to Bob's photon to turn it into a photon with the same state as x.

The special case in which $|\Psi\rangle_x = |\Psi\rangle_B$ is called "lucky" by Zeilinger's team. It happens whenever a photon is detected in each of Alice's SPCMs as reported by her coincidence counter, in which case we know that Bob's photon must have the same state as x without any need for further processing. As such, analyzing B through a PBS (Bob) should demonstrate that B has the same polarization as the one we gave x, proving successful teleportation.

For example, if we set the polarizer to vertical, x is vertical (although neither Alice nor Bob know it), and Bob's photon should always exit the PBS's vertical port whenever Alice's coincidence counter indicates the "lucky" case. On the other hand, if the polarizer is set to horizontal, Bob's photon should only exit through the PBS's horizontal port when Alice detects a "lucky" coincidence.

As you can see, this system was able to measure only one of the four possible Bell states, so it could only teleport those photons that totally by chance would produce the "lucky" state. Even under those conditions, the experiment wasn't perfect because it produced only 80% detections of the correct state on Bob's detectors. However, this was good enough to prove the concept, given that without teleportation happening, a detection rate of only 50% would have been expected.

One of the major issues for effective teleportation is that Alice's entangled photon (A) and the photon to be teleported (x) must reach the beam splitter at exactly the same time, because the Bell-state measurement requires that the photons not be identifiable from each other. Even a small difference in time of arrival would—in principle—act as a path labeler, identifying which photon (A or x) ended up at which detector. The incredibly short pump laser pulse is used to produce almost simultaneous photons, and their time of arrival can be equalized by making each photon travel the right distance before arriving at the beam splitter. However, even careful calibration of the paths was not sufficient to ensure simultaneous arrival of the photons, so a very clever trick suggested by Marek Zukowski of the University of Gdansk in Poland was incorporated by Zeilinger's team.

The idea is to pass photons A and x through very narrow band-pass filters that pin down the frequency of the photons very precisely. As you may remember from chapter 6, Heisenberg's Uncertainty Principle will smear the photons in time, making it easier to ensure that both photons reach the beam splitter simultaneously without any possibility of being identified by their time of arrival.

More recently, in 2004, Anton Zeilinger and his group at the University of Vienna, performed successful photon teleportation between laboratories located on two different sides of the river Danube at a distance of 600 m from each other.[58,49] The basic components of the system are shown in Figure 154. Besides the large separation between Alice and Bob, this experiment incorporated more comprehensive measurement of the Bell state, as well as an electro-optic modulator to rotate Bob's photon according to the results of Alice's Bell-state measurement.

The setup is conceptually similar to the one in Figure 153, but much of the photons' travel takes place inside optical fibers instead of free air. In fact, the critical beamsplitting for the Bell-state measurement is done in a type of all-fiber beam splitter built by thermally tapering and fusing two optic fibers so that their cores come into intimate contact. In addition, the entangled photon sent to Bob from the entangler travels via a long optical fiber laid along a utilities tunnel under the River Danube.

You may see that Alice now has four SPCMs looking at the outputs of two PBSs instead of just two SPCMs detecting photons coming directly out of the Bell-state measurement beam splitter. This configuration allows Alice to distinguish two of the four possible Bell states. Since light travels more slowly through fiber than the radio message travels through air, Bob has time to use the results of Alice's measurements to rotate the polarization of his photon. He does it with an electro-optic modulator, which contains a crystal that rotates the polarization as a function of applied voltage. If Alice tells Bob that $|\Psi\rangle_x = |\Psi\rangle_B$, his photon is already identical to the original state of x, and he doesn't apply any voltage to the electro-optic modulator.

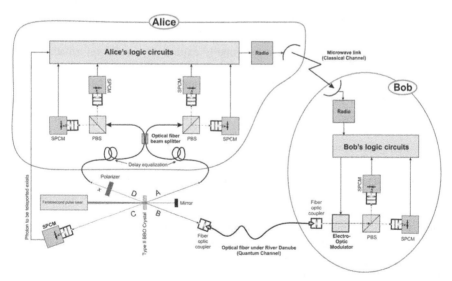

Figure 154   This setup was used by Anton Zeilinger's group at the University of Vienna to teleport photons at a distance of 600 m under the Danube River. Besides the large separation between Alice and Bob, this experiment incorporated an electro-optic modulator to rotate Bob's photon according to the results of Alice's Bell-state measurement.

On the other hand, if the result of the Bell-state measurement indicates that $|\Psi\rangle_x = |\Psi\rangle_B + 90°$, Bob applies a high voltage to the electro-optic modulator to rotate his photon by $90°$. This manages to teleport a maximum of 50% of the photons, instead of the 25% possible using only the "lucky" state.

Besides demonstrating that teleportation works over greater distances than on a lab breadboard, the most significant part of this experiment was that the fiber-optic lines were routed under real-world conditions—just as a telecommunications optical fiber would be. In the case of the Danube experiment, the optical fiber was sent through an underground sewage pipe tunnel, which exposed the fiber-optic link to temperature fluctuations, vibrations, and other environmental factors that could interfere with the process. This experiment was critical in establishing proof of concept for the nascent industry of quantum communications.

FASTER-THAN-LIGHT COMMUNICATION?

The fact that the state of Bob's photon changes instantaneously when Alice makes her Bell-state measurement is often confused by nonphysicists as a technology that could be used for communicating faster than light. The error is in thinking that an entangled photon encodes the information, when in fact the information is encoded in the *correlation* between the entangled photons.

Look back at Figure 152. Without the results of Alice's Bell-state measurement, Bob wouldn't know which one of the four corrections to apply to his photon to replicate x. Since these are completely random, Bob's attempt at guessing which correction to apply to every photon would succeed only 25% of the time, and he won't even know when he guessed correctly. His data stream would be as useless as if he had taken a beam of photons with completely random polarizations. Bob must simply wait for the classical message from Alice to properly extract any useful information, since the information is encoded in the correlation between Alice's and his measurements. Einstein's Relativity is not violated by the quantum teleporter, because even with "spooky action at a distance" instantaneously connecting the entangled photons, no information can be conveyed any faster than it takes for Alice's classical message to arrive to Bob.

Few people outside the physics community understand why superluminal communication is so problematic. After all, no major issues seem to be caused in *Star Trek* by the use of "subspace communications" to establish nearly instantaneous contact with people and places light-years away. However, being able to transfer information at speeds faster than light opens up the possibility of communicating with the past, which creates paradoxes that give real physicists real heartburn.

Imagine that Alice is at rest and Bob is traveling in a spaceship close to the speed of light c, such that Alice's clock appears to run twice as fast as Bob's.[¶¶¶] If Alice could send Bob a signal that traveled faster than c, it would appear to Bob that her

---

[¶¶¶]This requires an understanding of the Theory of Relativity, which is outside the scope of this book. However, you can read about it and return to this argument to understand how faster-than-light signaling leads to paradoxical situations epitomized by the "grandfather paradox" in which a time traveler goes back to the past and kills his grandfather, preventing his own birth.

signal was going backward in time. When Bob replied to Alice, his signal would reach Alice before she sent the original message. Thus, Alice could use Bob as a relay to send a message into her own past.

Any mechanism that would allow one to build a device capable of signaling to the past (commonly known as a "time telegraph") is usually taken by physicists as evidence that it is impossible, because it violates the order of cause and effect. For example, when physicist Nick Herbert proposed a quantum cloning machine that could copy a single photon exactly,[60] Israeli physicist Asher Peres prompted the physics community to look for the flaw in this proposal, since it would lead to the time telegraph paradox, indicating something had to be wrong in principle with this idea.

The argument is that if Bob had a quantum cloning machine, he could take the photon from the entangled pair that he shared with Alice and make some large number of perfect copies. He would then take all of his cloned photons (say 100 of them) and pass them through a vertical polarizer. Here comes the clincher: if Alice measures her photon with a vertical or a horizontal polarizer, either 0% or 100% of Bob's photons will go through the polarizer. However, if Alice measured her photon with a 45° polarizer, only 50% of his photons will make it through his polarizer. Therefore, Alice could instantaneously signal Bob by simply changing the angle of her polarizer.

In fact, within a few months, physicists Wojciech Zurek, William Wootters, and Dennis Dieks proved mathematically that it is impossible to clone quantum information, leading to the so-called Quantum No-Cloning Theorem.[61,62] Interestingly, the No-Cloning Theorem in itself doesn't prevent superluminal communication via quantum entanglement, but it does prevent it through the mechanism proposed by Nick Herbert, which had seemed promising.

QUANTUM CRYPTOGRAPHY

Coding messages was traditionally of interest only to diplomats, the military, and spy agencies. However, cryptography is now an essential part of everyone's life, given that the Internet is used to communicate all of our bank transactions, industrial designs, and commercial dealings.

Most systems today use an encrypting method known as *public-key cryptography*, in which Bob (the person who anticipates receiving messages) first creates both a public key and an associated private key, and publishes the public key. When Alice wants to send a secure message to Bob, she encrypts it using Bob's public key. To decrypt the message, Bob uses his private key.

Key generation is based on a calculation that is very easy for Bob to make, but very difficult to reverse by an eavesdropper—the e-e-evil Eve. For example, given two numbers, a computer can very quickly multiply them together to find the product, which is provided as the public key. However, it is much more difficult for an eavesdropper to find the factors given only the product. Nevertheless, the weakness of current public-key cryptography is that a powerful computer (for example, a quantum computer, as we will soon see) could use the public key to learn the private key. As such, key distribution systems based on this idea are very practical and efficient, but their security is based on the assumption that the eavesdropper does not have access to massive computing power.

Once Eve the eavesdropper uses the supercomputer in her lair to crack the private key, she can tap the encrypted channel (e.g., a radio or Internet transmission) and read the secret messages without either Alice or Bob knowing their communications have been compromised.

Traditionally, the only way to ensure completely private communications was to use a long random key shared securely between the parties and used only once. In theory, using such *one-time pad* makes the message secure, but there is still the problem that the key, which has to be distributed, may be susceptible to interception. In addition, reusing a one-time pad allows sophisticated code-breakers to find patterns that can reveal the key.

A new way of ensuring that even a sophisticated eavesdropper cannot decode an encrypted message is to distribute a key using a quantum channel that cannot be intercepted without being detected by the sender and receiver. This is possible because the act of measuring a quantum state will cause changes that can be detected. Putting it simply, when Bob and Alice exchange a key sent via a quantum channel, they can spot Eve's tampering, because her measurements of the photon stream will cause detectable errors in the data. As such, Bob and Alice only send an encrypted message using a key that is known to be secure.

A method that is now being employed commercially for *quantum key distribution* (or *QKD*) was first proposed by Charles Bennett of IBM and Gilles Brassard of the University of Montreal[63]—the same brilliant guys who came up with the protocol for quantum teleportation. Using their procedure, which is commonly known as BB84, Alice generates a stream of individual photons polarized in one of two modes (also known as basis): vertical/horizontal, or diagonally $\pm 45°$. Within each mode, one orientation represents a digital "0" and the other a "1." Alice randomly chooses both a mode (polarization frame) and an orientation (digital value) for each photon sent over the quantum channel. Depending on her random selections, the photons she sends to Bob have the polarizations shown in Table 12.

Alice records the mode, digital value, and exact time of transmission for each photon she sends to Bob. As the receiver, Bob randomly chooses between the two modes when he tries to detect a photon. If he chances to choose the same mode that Alice used for a given photon, he will correctly measure its orientation and determine its digital value. Choosing a different mode from the one Alice used will give him the wrong value for that photon. However, he doesn't know which measurement is right or wrong. Bob also records the mode he used, result he obtained, and exact time of arrival for each photon.

TABLE 12   Photon Polarizations Used over the Quantum Channel in the BB84 Quantum Cryptography Protocol

		Digital value	
Mode (basis)		0	1
Rectilinear (0° and 90°)	+	↑	→
Diagonal (45° and 135°)	×	↗	↘

TABLE 13   Example of How a One-Time-Pad Encryption Key is Generated by Alice and Bob Using the BB84 Quantum Cryptography Protocol

Photon sent by Alice	1	2	3	4	5	6	7
Alice's random bit selection	0	0	1	0	1	1	0
Alice's random mode selection	+	×	×	×	+	×	×
Polarization of photons sent by Alice	↑	↗	↘	↑	→	↘	↗
Bob's random measurement mode selection	×	×	+	×	+	+	×
Photon polarization measured by Bob	↗	↗	→	↑	→	→	↗
Shared secret key		0		0	1		0

Alice uses a classical channel to tell Bob the mode she used for encoding each photon, but does not tell him its digital value. Bob then ignores all instances where he measured a photon in the wrong mode, and tells Alice which ones he measured correctly, not telling her the digital value he measured. Alice then discards all the photons Bob didn't measure correctly. Only the set of photons measured correctly by Bob are assembled into the one-time-pad encryption key. Table 13 and Figure 155 show an example of the process.

Let's now suppose that Eve attempts to eavesdrop on Bob and Alice (Figure 156). She will need to intercept Alice's photon stream with her detector. However, like Bob, Eve doesn't know the modes used by Alice, so she needs to measure at random modes. Most importantly, she destroys Alice's photons when she measures them, so she must generate a new quantum message to send to Bob to disguise her prying. Eve has to guess the polarization of many of the photons, which creates errors in the string of values used in the encryption key. Bob and

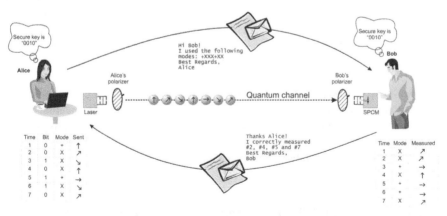

Figure 155   To generate a one-time-pad BB84 encryption key, Alice first sends Bob a sequence of photons with polarizations encoding a set of random bits and modes. Bob measures these photons at a set of random modes. Alice then tells Bob the modes she used (but not her digital values), and Bob replies with a list of photons for which he, by chance, measured using the same mode as Alice. The bits encoded in the matching photons are used to compose the encryption key.

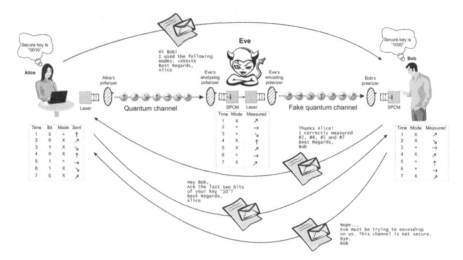

Figure 156   If Eve the eavesdropper tries to measure and resend the photons, errors occur in the data, so Alice's key and Bob's key don't match. Alice and Bob are alerted about Eve's presence when they compare just a few bits of the key.

Alice can easily find these errors by comparing a subset of their remaining bit strings, telling them there is an eavesdropper and they must therefore discard the key.

There are other QKD protocols, including some that use entangled pairs to further boost key security. QKD products are already commercially available from id Quantique (Switzerland), MagiQ Technologies (U. S.), SmartQuantum (France), and Quintessence Labs (Australia).

One last word about QKD. Although quantum cryptography is theoretically secure, the real-world components used to implement it are vulnerable to attacks. In 2010, a team from the Norwegian University of Science and Technology in Trondheim was able to successfully eavesdrop the secret key without leaving a trace by exploiting some characteristics of the SPAD-based SPCMs in the commercial QKD system made by id Quantique. The hack involves Eve flashing laser pulses on Bob's SPCMs to reduce their sensitivity using a clever pulsing method[64] that allows Eve to safely intercept a message without leaving the telltale quantum errors. This vulnerability can be closed in a number of ways, including redesigning the SPCMs, but it reminds us that no real lock is completely safe, not even one that is based on a theoretically unbreakable mechanism.

QUANTUM COMPUTING AND TECHNOLOGIES FOR THE FUTURE

We had mentioned earlier that quantum information is very different than classical information. We saw that while classical information is independent of its physical representation, quantum information cannot be read without disturbing the physical

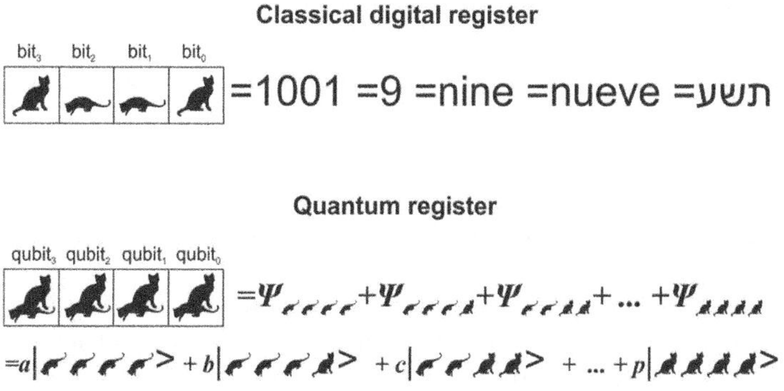

Classical digital register

$=1001 =9 =$nine $=$nueve $=$תשע

Quantum register

$$=\Psi_{\text{卂卂卂卂}} + \Psi_{\text{卂卂卂卂}} + \Psi_{\text{卂卂卂卂}} + \ldots + \Psi_{\text{卂卂卂卂}}$$

$$=a|\text{卂卂卂卂}> + b|\text{卂卂卂卂}> + c|\text{卂卂卂卂}> + \ldots + p|\text{卂卂卂卂}>$$

Figure 157   A classical 4-bit register can store only one of 16 different numbers from zero = 0000 to fifteen = 1111. It doesn't matter what the method of representation—dead or live cats, ones or zeros, decimal numbers, or words—the classical information doesn't change. On the contrary, a 4-qubit quantum register stores a superposition of all 16 different numbers at the same time. The coefficients $a, b, c, \ldots p$ are related to the probability of each state. In a quantum computer, all 16 states can be manipulated at once, allowing much faster calculation than with a classical computer.

process encoding it. In addition, we mentioned that the Quantum No-Cloning Theorem prevents us from copying quantum information without errors.

However, there is a third difference between quantum and classical information that opens up incredible possibilities for the future. The concept is that all possible data that could be held by a quantum system are simultaneously contained by that quantum system. For example, one qubit simultaneously contains a digital "0" and a digital "1" in a superposition of states. Two qubits simultaneously hold "00", "01", "10," and "11" in quantum superposition. As shown in Figure 157, the number of states allowed in the superposition grows exponentially with the number of available qubits. For an n-qubit register, the number of states in superposition is 2^n. For a modest 500-qubit register, the number of states in superposition can be 2^{500}, a number that vastly exceeds the number of atoms in the whole universe!

As we saw in the analysis of teleportation, quantum states can be manipulated without destroying the quantum superposition, which makes it possible to compute with quantum information. The advantage of quantum computing over classical computing is that each calculation simultaneously evaluates all of the possibilities that can be encoded by the system. A single computational operation on a quantum system containing n qubits is the equivalent of performing 2^n operations on a classical computer.

Let's try to explain this difficult concept in a simpler way. As shown in Figure 157, let's suppose that we have a 4-qubit register made of four Schrödinger's cat boxes. If we finally open the boxes (we make a measurement), the register will indicate a certain 4-bit number in which each bit will be 卂 $= 1$, 卂 $= 0$. However, while the boxes are sealed, each cat is in a superposition of dead and alive states. As such, the register is in a superposition of all of its probable states:

$$\Psi_{4-\text{qubit register}} = \Psi_{\text{卂卂卂卂}} + \Psi_{\text{卂卂卂卂}} + \Psi_{\text{卂卂卂卂}} + \ldots + \Psi_{\text{卂卂卂卂}}$$

We could put this in a bra-ket notation:

$$|\Psi_{4-\text{qubit register}}> \, = a|🐀🐀🐀🐀> + b|🐀🐀🐀🐁> + c|🐀🐀🐁🐁> + \ldots$$
$$+ p|🐁🐁🐁🐁>$$

where the variables a, b, c, $\ldots$ p are related to the probability of each state. These numbers could be positive, negative, or complex, as long as the total probability adds up to one.

The power of a quantum computer is realized if we find a sequence of operations that can modify the quantum state (the value of the individual probabilities) without destroying the quantum superposition. Each operation modifies all of the quantum states at once, so a single quantum operation performs the equivalent of 2^n classical operations. These operations are the quantum computer's algorithm, and the result is obtained by finally making a measurement and collapsing the 4 qubits into 4 classical bits.

The power of a quantum computer scales exponentially as the number of qubits grows. For example, a modest quantum computer containing just 500 qubits would be able to perform in a single operation the gargantuan task of conducting 2^{500} classical operations—something that not even the largest classical, massively parallel supercomputer can do today with a single operation!

All of this is largely theoretical at the moment. The hardware for making a quantum computer containing more than just a few qubits hasn't been designed yet, and very few quantum algorithms have been invented. The whole field of quantum information is certainly in its earliest infancy. Hopefully, this book will inspire you to be one of the pioneers of the coming quantum information era.

EXPERIMENTS AND QUESTIONS

1. What are the concepts of locality, realism, and completeness as used by Einstein, Podolsky, and Rosen (EPR)?

2. Explain what the EPR paper tried to show. How does the EPR argument conflict with the Copenhagen Interpretation?

3. How does Bell's Inequality determine whether or not a local hidden-variables theory underlies physical reality? Explain how coincidence-counting statistics are used experimentally to test for violations of Bell's inequality.

4. Construct and align the entangled-photon setup shown in Figure 148. Count the number of individual photons and coincidences during 15 s when the polarizers are set to $\alpha = \beta = 0°$, $\alpha = \beta = 45°$, $\alpha = \beta = 90°$, $\alpha = 0°$ and $\beta = 90°$. What do these photon statistics tell you?

5. Why is the UV pump laser beam polarized at 45° in the entangled-photon source of Figure 142?

6. Collect photon and coincidence-count data at the polarization angles shown in Table 10. Calculate the 16 correlations $E(\alpha, \beta)$ and test whether or not they violate the CHSH form

of the Bell Inequality. What value of S did you obtain? What do your results demonstrate? What would it mean if $|S| \leq 2$?

7. What loopholes are left open by this experiment to determine the violation of Bell's inequalities? How could these loopholes be closed?

8. Why is superluminal communication problematic to physicists? How can information be encoded using entangled photons? Why doesn't entanglement violate the Theory of Relativity?

9. In what way does quantum teleportation process quantum information? Why is quantum teleportation important to quantum computing?

10. Why does the Copenhagen Interpretation prohibit the cloning of quantum states? What physical paradoxes could occur if quantum cloning would be possible?

11. What is the chief advantage of encryption schemes based on quantum key distribution over conventional cryptography methods?

12. Why does the development of a practical quantum computer pose a threat to classical encryption methods?

REFERENCES

1. Curry SM, Schawlow AL, Measuring the Diameter of a Hair by Diffraction, *American Journal of Physics* **42**(5), 412–413, 1974.
2. Brody J, Griffin L, Segre P, Measurement of the Speed of Light in Water Using Foucault's Technique, *American Journal of Physics* **78**(6), 650–653, 2010.
3. Petersen DN, The Care and Feeding of Gunnplexers, *QST* 14–18, April 1983.
4. Richardson B, *The Gunnplexer Cookbook*, Ham Radio Publishing Group, Greenville, 1981.
5. Garver WP, The Photoelectric Effect Using LEDs as Light Sources, *The Physics Teacher* **44**, 272–275, 2006.
6. Rudnick J, Tannhauser DS, Concerning a Widespread Error in the Description of the Photoelectric Effect, *American Journal of Physics* **44**(8), 796–798, 1976.
7. Spectrum Techniques, *ST400 Scintillation Processor Operating Manual*, Spectrum Techniques, Oak Ridge, TN, 2001; www.spectrumtechniques.com/manuals/ST400manual.pdf.
8. Hansen SP, Reconfigurable Glass Vacuum Chambers, *The Bell Jar* **2**, 20–23, 1993.
9. Hansen SP, A Kit of Components for Conducting Gaseous Discharge and Electron Beam Experiments, *The Bell Jar* **2**, 28–30, 1993.
10. Perrin J, New Experiments on the Kathode Rays, Read before the Paris Academy of Sciences, December 30, 1895. Translation published in *Nature* **53**, 298–299, 1896, Reprinted in *The Bell Jar* **9**(2), 8–9, 2000.
11. Thomson JJ, Cathode Rays, *Philosophical Magazine* **44**, 293, 1897.
12. Calvert JB, *The Cathode-Ray Tube*, 2002; http://mysite.du.edu/~etuttle/electron/elect29.htm.
13. Hansen SP, A Functional Thomson *e/m* Apparatus, *The Bell Jar* **6**(3/4), 5–7, 1996.
14. Thomson JJ, On the Structure of the Atom: An Investigation of the Stability and Periods of Oscillation of a Number of Corpuscles Arranged at Equal Intervals Around the Circumference of a Circle; with Application of the Results to the Theory of Atomic Structure, *Philosophical Magazine* Series 6, **7**(39), 237–265, 1904.
15. Hoag JB, *Electron and Nuclear Physics*, D. Van Nostrand, New York, 1929.
16. Connell LF, Use of a Cathode-Ray Oscilloscope in Hoag's *e/m* Experiment, *American Journal of Physics* **17**(4), 222–223, 1949.
17. Yang YK, Determining the Ratio of Charge to Mass *e/m* for Electrons by Magnetic Focusing, *American Journal of Physics* **66**(2), 157–162, 1998.
18. Dowel G, *Converting the ENi CD V-700 model 6-b to LENi*, http://tech.groups.yahoo.com/group/GeigerCounterEnthusiasts/files/Converting%20ENi%206b%20to%20LENi/.
19. Thomson C, *A Low-Cost "Paint Can" Scintillator*, 2002; http://cdtsys.com/PaintCanScint.html.
20. Rutherford E, The Scattering of Alpha and Beta Particles by Matter and the Structure of the Atom, *Philosophical Magazine* **21**, 669–688, 1911.
21. Klein AG, Caspar MJ, Nicola A, Simple Alpha Particle Detector, *American Journal of Physics* **53**(12), 1212, 1985.
22. Jönsson G, Alpha Particle Detection by Plastic Film—An Experimental Approach to Nuclear Physics, *Physics Education* **16**(2), 102–103, 1981.
23. Bertsch GF, Nolen JA, Simple Demonstration of the Compton Effect, *American Journal of Physics* **52**(2), 183–184, 1984.

Exploring Quantum Physics Through Hands-On Projects. By David Prutchi and Shanni R. Prutchi
© 2012 John Wiley & Sons, Inc. Published 2012 by John Wiley & Sons, Inc.

24. Mudhole TS, Umakantha N, Determination of the Rest-Mass Energy of the Electron: A Laboratory Experiment, *American Journal of Physics* **45**(11), 1119–1120, 1977.

25. Merli PG, Missiroli GF and Pozzi G, On the Statistical Aspect of Electron Interference Phenomena, *American Journal of Physics* **44**(3), 306–307, 1976.

26. Tonomura A, Endo J, Matsuda T, Kawasaki T and Ezawa H, Demonstration of Single-Electron Buildup of an Interference Pattern, *American Journal of Physics* **57**, 117–120, 1989.

27. Panitz JA, Rempfer G, A Transmission Electron Microscope for Lecture Demonstrations, *American Journal of Physics* **74**(11), 953–956, 2006.

28. Stong CL, The Amateur Scientist—A High School Physics Club Builds Electron Microscopes, *Scientific American* September 1973, 184–189, **229**(3).

29. Zeilinger A, Horne MA, Aharonov-Bohm with Neutrons, *Physics World 2* **23**, 1989.

30. Carnal O, Mlynek J, Young's Double-Slit Experiment with Atoms: A Simple Atom Interferometer, *Physical Review Letters* **66**(21), 2689–2692, 1991.

31. Nairz O, Arndt M, Zeilinger A, Experimental Verification of the Heisenberg Uncertainty Principle for Fullerene Molecules, *Physical Review A* **65**(3), 2002; DOI: 10.1103/PhysRevA.65.032109.

32. Hackermüller L, Uttenthaler S, Hornberger K, Reiger E, Brezger B, Zeilinger A, Arndt M, Wave Nature of Biomolecules and Fluorofullerenes, *Physics Review Letters* **91**, 90408, 2003.

33. Trafton A, Supercooling May Yield View of Quantum Effects, *MIT TechTalk* **51**(23), 4, 2007.

34. Boatman EM, Lisensky GC, Nordell KJ, A Safer, Easier, Faster Synthesis for CdSe Quantum Dot Nanocrystals, *Journal of Chemical Education* **82**, 1697–1699, 2005.

35. Lisensky G, Preparation of Cadmium Selenide Quantum Dot Nanoparticles, 2010; http://mrsec.wisc.edu/Edetc/nanolab/CdSe/index.html.

36. Albiol F, Navas S, Andres MV, Microwave Experiments on Electromagnetic Evanescent Waves and Tunneling Effect, *American Journal of Physics* **61**(2), 165–169, 1993.

37. Nimtz G, Stahlhofen AA, Macroscopic Violation of Special Relativity, *arXiv* August 5, 2007; arXiv:0708.0681v1.

38. Hillmer R, Kwiat P, A Do-It-Yourself Quantum Eraser, *Scientific American* May 2007, 90–95.

39. Dürr S, Nonn T, Rempe G, Origin of Quantum-Mechanical Complementarity Probed by a "Which-Way" Experiment in an Atom Interferometer, *Nature* **395**, 33–37, 1998.

40. Einstein A, Podolsky B, Rosen N, Can Quantum-Mechanical Description of Physical Reality Be Considered Complete? *Physics Review* **47**, 777–780; DOI:10.1103/PhysRev.47.777.

41. Chiao RY, Kwiat PG, Steinberg AM, Faster than Light?, *Scientific American* August 1993, 52–60.

42. Mermin ND, Is the Moon There When Nobody Looks? Reality and the Quantum Theory, *Physics Today* **38**(4), 38–47, 1985.

43. Mermin ND, Bringing Home the Atomic World: Quantum Mysteries for Anybody, *American Journal of Physics* **49**(10), 940–943, 1981.

44. Mermin ND, Quantum Mysteries Revisited, *American Journal of Physics* **58**(8), 731–734, 1990.

45. Dehlinger D, Mitchell MW, Entangled Photons, Nonlocality, and Bell Inequalities in the Undergraduate Laboratory, *American Journal of Physics* **70**(9), 903–910, 2002.

46. Kwiat PG, Waks E, White AG, Appelbaum I, Eberhard PH, Ultra-Bright Source of Polarization-Entangled Photons, *Physical Review A* **60**(2), 773–776, 1999.

47. Thorn JJ, Neel MS, Donato VW, Bergreen GS, Davies RE, Beck M, Observing the Quantum Behavior of Light in an Undergraduate Laboratory, *American Journal of Physics* **72**(9), 1210–1219, 2004.

48. Dehlinger D, Mitchell MW, Entangled Photon Apparatus for the Undergraduate Laboratory, *American Journal of Physics* **70**(9), 898–902, 2002.

49. Zeilinger A, *Dance of the Photons: From Einstein to Quantum Teleportation*, Farrar, Straus and Giroux, New York, 2010.

50. Aspect A, Grangier P, Roger G, Experimental Realization of Einstein–Podolsky–Rosen–Bohm *Gedankenexperiment*: A New Violation of Bell's Inequalities, *Physical Review Letters* **49**(2), 91–94, 1982.

51. Aspect A, Dalibard J, Grangier P, Roger G, Quantum Beats in Continuously Excited Atomic Cascades, *Optics Communications* **49**(6), 429–434, 1984.

52. Clauser JF, Horne MA, Shimony A, Holt RA, Proposed Experiment to Test Local Hidden-Variable Theories, *Physical Review Letters* **23**, 880–884, 1969.

53. Rowe MA, Kielpinski D, Meyer V, Sackett CA, Itano WM, Monroe C, Wineland DJ, Experimental Violation of a Bell's Inequality with Efficient Detection, *Nature* **409**, 791–794, 2001.
54. Aspect A, Dalibard J, Roger G, Experimental Test of Bell's Inequalities Using Time-Varying Analyzers, *Physical Review Letters* **49**(25), 1804–1807, 1982.
55. Tittel W, Brendel J, Zbinden H, Gisin N, Violation of Bell Inequalities by Photons More Than 10 km Apart, *Physics Review Letters* **81**(17), 3563–3566, 1998.
56. Jennewein T, Achleitner U, Weihs G, Weinfurter H, Zeilinger A, A Fast and Compact Quantum Random Number Generator, *Review of Scientific Instrumentation* **71**(4), 1675–1680, 2000.
57. Bennett CH, Brassard G, Crépeau C, Jozsa R, Peres A, Wootters WK, Teleporting an Unknown Quantum State via Dual Classical and Einstein–Podolsky–Rosen Channels, *Physical Review Letters* **70**(13), 1895–1899, 1993.
58. Ursin R, Jennewein T, Aspelmeyer M, Kaltenbaek R, Lindenthal M, Walther P, Zeilinger A, Quantum Teleportation Across the Danube, *Nature* **430**(7002), 849, 2004.
59. Walker J, The Amateur Scientist—A Homemade Device for Testing Particle Scattering, *Scientific American* **254**(2), 114–118.
60. Herbert N, FLASH—A Superluminal Communicator Based upon a New Type of Quantum Measurement, *Foundations of Physics* **12**(2), 1171, 1982.
61. Dieks D, Communication by EPR Devices, *Physics Letters A* **92**(6), 271–272, 1982.
62. Wootters WK, Zurek WH, A Single Quantum Cannot Be Cloned, *Nature* **299**(5886), 802–803, 1982.
63. Bennett CH, Brassard G, Quantum Cryptography: Public Key Distribution and Coin Tossing, in *Proceedings of the 1984 IEEE International Conference on Computers, Systems, and Signal Processing, Bangalore*, 175, IEEE, Piscataway, 1984.
64. Lydersen L, Wiechers C, Wittmann C, Elser D, Skaar J, Makarov V, Hacking Commercial Quantum Cryptography Systems by Tailored Bright Illumination, *Nature Photonics* **4**, 686–689, 2010.
65. Jan O, Makotyn P, *Final Report—Single Photon Avalanche Diode Module*, ECE445 Senior Design Project, University of Illinois at Urbana–Champaign, 2006. Available online at: http://courses.engr.illinois.edu/ece445/projects/summer2006/project1_final_paper.doc.
66. Nairz O, Arndt M, Zeilinger A, Quantum Interference Experiments with Large Molecules, *American Journal of Physics* **71**(4), 319–325, 2003.
67. Weis A, Wynands R, Three Demonstration Experiments on the Wave and Particle Nature of Light, *Physik und Didaktik in Schule und Hochschule* **1**(2), 67–73, 2003.
68. Zwiller V, Blom H, Jonsson P, Panev N, Jeppesen S, Tsegaye T, Goobar E, Pistoll ME, Samuelson L, Björk G, Single Quantum Dots Emit Single Photons at a Time: Antibunching Experiments, *Applied Physics Letters* **78**, 2476–2478, 2001.

SOURCES FOR MATERIALS AND COMPONENTS

Ace Glass
www.aceglass.com
1430 North West Boulevard
Vineland, NJ 08362
(856) 692-3333

Advanced Receiver Research (also known as Ar^2 Communications Products)
www.advancedreceiver.com
Box 1242
Burlington, CT 06013
(860) 485-0310

Anchor Optics
www.anchoroptics.com
101 East Gloucester Pike
Barrington, NJ 08007
(856) 573-6865

Apex Electronics
www.apexelectronic.com
8909 San Fernando Road
Sun Valley, CA 91352
(818) 767-7202

Army Radio Sales Co.
www.armyradio.co.uk
109 Booth Road, Colindale
London NW9 5JU
United Kingdom
+44 (0)7905-671741

Exploring Quantum Physics Through Hands-On Projects. By David Prutchi and Shanni R. Prutchi
© 2012 John Wiley & Sons, Inc. Published 2012 by John Wiley & Sons, Inc.

The Bell Jar
www.belljar.net
c/o DiverseArts
34 Meadowbrook Lane
Owl's Head, ME 04854

The Black Hole of Los Alamos
www.blackholesurplus.com
4015 Arkansas Road
Los Alamos, NM 87544
(505) 662-5053

CENCO Physics
(A division of Sargent-Welch)
www.cencophysics.com
P.O. Box 4130
Buffalo, NY 14217
(800) 727-4368

Edmund Optics
www.edmundoptics.com
101 East Gloucester Pike
Barrington, NJ 08007
(800) 363-1992

Evident Technologies
www.evidenttech.com
45 Ferry Street
Troy, NY 12180
(518) 273-6266

Fair Radio Sales
www.fairradio.com
2395 St. Johns Road
Lima, OH 45804
(419) 223-2196

George Dowell
eBay® Seller: GeoElectronics
56791 Rivere Au Sel Place
New London, MO 63459

Images Scientific Instruments, Inc.
www.imagesco.com
109 Woods of Arden Road
Staten Island, NY 10312
(718) 966-3694

Information Unlimited
www.amazing1.com
PO Box 716
Amherst, NH 03031
(603) 673-6493

Kurt J. Lesker Company
www.lesker.com
1925 Route 51
Clairton, PA 15025
(412) 387-9200

Landauer
www.landauer.com
2 Science Road
Glenwood, IL 60425
(708) 755-7000

Laser Components IG, Inc.
www.lasercomponents.com
9 River Road
Hudson, NH 03051
(603) 821-7040

Matsusada Precision, Inc.
www.matsusada.com
80 Orville Drive, Suite 100
Bohemia, NY 11716
(631) 244-1407

McMaster-Carr
www.mcmaster.com
600 North County Line Road
Elmhurst, IL 60126
(630) 833-5375

Micro-Mark
www.micromark.com
340 Snyder Avenue
Berkeley Heights, NJ 07922
(908) 464-2984

Murphy's Surplus Warehouse
www.murphyjunk.bizland.com
401 North Johnson Avenue
El Cajon, CA 92020
(619) 444-7717

Nanosys, Inc.
www.nanosysinc.com
2625 Hanover Street
Palo Alto, CA 94304
(650) 331-2100

National Instruments
www.ni.com
11500 North Mopac Expressway
Austin, TX 78759
(800) 531-5066

Newlight Photonics, Inc.
www.newlightphotonics.com
772 Dovercourt Road
Toronto, Ontario
Canada M6H 3A5
(416) 536-8368

Newport
www.newport.com
1791 Deere Avenue
Irvine CA 92606
(949) 863-3144

Parts Express
www.parts-express.com
725 Pleasant Valley Drive
Springboro, OH 45066
(800) 338-0531

PerkinElmer
www.perkinelmer.com
940 Winter Street
Waltham, MA 02451
(781) 663-6900

Photop Technologies, Inc.
www.photoptech.com
21949 Plummer Street
Chatsworth, CA 91311
(818) 678-1999

Sargent-Welch
www.sargentwelch.com
777 East Park Dr.
Tonawanda, NY 14150
(800)727-4368

Science First
www.sciencefirst.com
86475 Gene Lasserre Boulevard
Yulee, FL 32097
(904) 225-5558

The Science Source
www.thesciencesource.com
299 Atlantic Highway
Waldoboro, ME 04572
(207) 832-6344

Skycraft Parts & Surplus
www.skycraftsurplus.com
2245 West Fairbanks Avenue
Winter Park, FL 32789
(407) 628-5634

Spectrum Scientifics
www.spectrum-scientifics.com
4403 Main Street
Philadelphia, PA 19127
(215) 667-8309

Spectrum Techniques
www.spectrumtechniques.com
106 Union Valley Road
Oak Ridge, TN 37830
(865) 482-9937

Sphere Research Corporation
www.sphere.bc.ca
3394 Sunnyside Road
West Kelowna, BC
Canada V1Z 2V4
(250) 769-1834

Structure Probe, Inc.
www.2spi.com
569 East Gay Street
West Chester, PA 19380
(610) 436-5400

Surplus Sales of Nebraska
www.surplussales.com
1218 Nicholas Street
Omaha, NE 68102
(402) 346-4750

Surplus Shed
www.surplusshed.com
1050 Maidencreek Road
Fleetwood, PA 19522
(610) 926-9226

TEL-Atomic, Inc.
www.telatomic.com
1223 Greenwood Avenue
Jackson, MI 49203
(517) 783-3039

Thorlabs
www.thorlabs.com
435 Route 206 North
Newton, NJ 07860
(973) 579-7227

United Nuclear Scientific
www.unitednuclear.com
239 East Grand River Road
Laingsburg, MI 48848
(517) 651-5635

ABBREVIATIONS

AC	alternating current
AWG	American wire gauge
BBO	β-barium borate
CCD	charge-coupled device
CCFL	cold-cathode fluorescent lamp
CHSH Test	Clauser, Horne, Shimony, and Holt Test
CRT	cathode-ray tube
DC	direct current
DMM	digital multimeter
DVD	digital versatile disk
EL	electroluminescent
EPR	Einstein, Podolsky, and Rosen
FFT	fast Fourier transform
FIP	female iron pipe
FWHM	full width at half-maximum
GM	Geiger–Müller
HeNe	helium–neon
IC	integrated circuit
ICCD	intensified CCD
ID	inner diameter
IR	infrared
ISIT	intensified silicon-intensified target
IST	intensified silicon target
KF	Klein flange
LBO	lithium triborate
LED	light-emitting diode
MCA	multichannel analyzer
MCP	microchannel plate
MSDS	material safety data sheet
NaI(Tl)	sodium iodide doped with thallium
NASA	National Aeronautics and Space Administration
OD	outer diameter
op amp	operational amplifier
PBS	polarizing beamsplitter

Exploring Quantum Physics Through Hands-On Projects. By David Prutchi and Shanni R. Prutchi
© 2012 John Wiley & Sons, Inc. Published 2012 by John Wiley & Sons, Inc.

PC	personal computer
PET	positron emission tomography
PMT	photomultiplier tube
PVC	polyvinyl chloride
QF	quick flange
QKD	quantum key distribution
RAM	random-access memory
RTV silicone	room temperature vulcanization silicone
SAE	Society of Automotive Engineers
SPAD	single-photon avalanche diode
SPCM	single photon–counting module
T/C	thermocouple
TEC	thermoelectric cooler
TEM	transmission electron microscope
TPI	turns per inch
TTL	transistor–transistor logic
USB	universal serial bus
UV	ultraviolet
VAC	volts, alternating current
VDC	volts, direct current

INDEX

Exploring Quantum Physics Through Hands-On Projects. By David Prutchi and Shanni R. Prutchi
© 2012 John Wiley & Sons, Inc. Published 2012 by John Wiley & Sons, Inc.

Printed and bound by CPI Group (UK) Ltd, Croydon, CR0 4YY

27/10/2024

14580272-0005